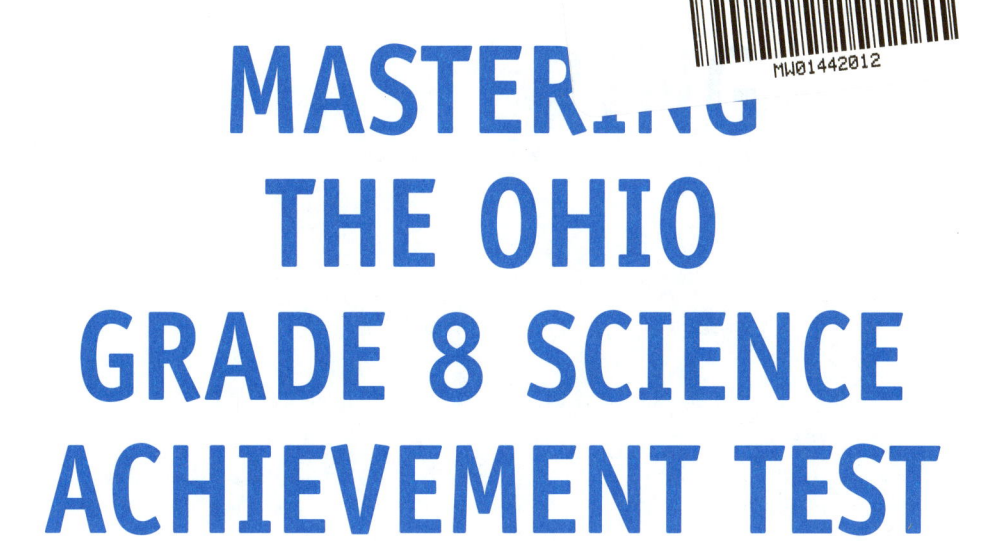

MASTERING THE OHIO GRADE 8 SCIENCE ACHIEVEMENT TEST

MARK JARRETT

Ph.D., Stanford University

STUART ZIMMER

JAMES KILLORAN

JARRETT PUBLISHING COMPANY

EAST COAST OFFICE
P.O. Box 1460
Ronkonkoma, NY 11779
631-981-4248

SOUTHERN OFFICE
50 Nettles Boulevard
Jensen Beach, FL 34957
800-859-7679

WEST COAST OFFICE
10 Folin Lane
Lafayette, CA 94549
925-906-9742

www.jarrettpub.com
1-800-859-7679   Fax: 631-588-4722

Copyright 2009 by Jarrett Publishing Company

Cover photo: Superstock, Inc.

All rights reserved. No part of this book may be reproduced in any form or by any means, including electronic, photographic, mechanical, or by any device for storage and retrieval of information, without the express written permission of the publisher. Requests for permission to copy any part of this book should be mailed to:

Jarrett Publishing Company
Post Office Box 1460
Ronkonkoma, New York 11779

ISBN 0-9795493-0-2 / 978-0-9795493-0-4
Printed in the United States of America
First Edition
10 9 8 7 6 5 4   09

ACKNOWLEDGMENTS

The authors would like to thank the following Ohio educators who helped review the manuscript. Their respective comments, suggestions, and recommendations have proved invaluable in preparing this book.

Dara Van Engelenhoven
Science Teacher
Columbus Public Schools
Columbus, Ohio

Treva E. Jeffries
Director, Science and Technology Education
Toledo Public Schools
Toledo, Ohio

Elizabeth Voit
Science Curriculum Director
Cleveland Public Schools
Cleveland, Ohio

Missi Zender, Ph.D
Science Curriculum Specialist
Summit County Educational Service Center
Cuyahoga Falls, Ohio

This book includes material from many different sources. Occasionally it is not possible to determine if a particular source is copyrighted, and if so, who is the copyright owner. If there has been a copyright infringement with any material produced in this book, it has been unintentional. We extend our sincerest apologies and would be happy to make immediate and appropriate restitution upon proof of copyright ownership.

Layout, graphics, and typesetting: Burmar Technical Corporation, Albertson, N.Y.

This book is dedicated…

to my wife, Gośka, and my children Alexander and Julia — *Mark Jarrett*

to my wife Joan, my children Todd and Ronald, and my grandchildren Jared and Katie Rose — *Stuart Zimmer*

to my wife Donna, my children Christian, Carrie, and Jesse, and my grandchildren Aiden, Christian, and Olivia — *James Killoran*

TABLE OF CONTENTS

UNIT 1: AN INTRODUCTION TO THE OHIO GRADE 8 SCIENCE ACHIEVEMENT TEST

★ **Chapter 1:** Ohio's Science Standards 2
★ **Chapter 2:** How to Answer Multiple-Choice Questions 5
★ **Chapter 3:** How to Answer Short- and Extended-Response Questions 15

UNIT 2: THE NATURE OF SCIENCE

★ **Chapter 4:** Scientific Ways of Knowing 25
★ **Chapter 5:** Scientific Inquiry 40
★ **Chapter 6:** Science and Technology 64

UNIT 3: THE PHYSICAL SCIENCES

★ **Chapter 7:** Matter 84
★ **Chapter 8:** Motion and Force 100
★ **Chapter 9:** Energy 107

UNIT 4: LIFE SCIENCES

★ **Chapter 10:** From Cells to Multicellular Organisms 128
★ **Chapter 11:** Reproduction and Inherited Traits 143
★ **Chapter 12:** Ecosystems and Adaptation to Ecological Change 153

UNIT 5: EARTH AND SPACE SCIENCES

★ **Chapter 13:** The Universe 175
★ **Chapter 14:** Earth's Interior and Landforms 192
★ **Chapter 15:** The Interaction of Earth's Systems 207

UNIT 6: A PRACTICE ACHIEVEMENT TEST IN SCIENCE

★ **A Practice Ohio Grade 8 Science Achievement Test** 223

    Glossary/Index 233

UNIT 1: AN INTRODUCTION TO THE OHIO GRADE 8 SCIENCE ACHIEVEMENT TEST

This year you will take the **Grade 8 Science Achievement Test**. Everyone wants to get a high score on this test. Unfortunately, just wanting a high score is not enough. You will really have to work at it! With this book as your guide, you should be better prepared for the test — and even enjoy studying for it!

WHAT WILL BE ON THE TEST?

Let's start by learning about the test itself. The **Ohio Grade 8 Science Achievement Test** has a total of **48 points**. There are three types of questions:

- Multiple-Choice Questions
- Short-Answer Questions
- Extended-Response Questions

MULTIPLE-CHOICE QUESTIONS
The test has 32 multiple-choice questions that count towards your final score. Each multiple-choice question will have four choices and be worth **1 point**. There will be four additional practice questions that **do not** count towards your final score.

SHORT-ANSWER QUESTIONS
Short-answer questions require a brief response in which you write a few sentences to answer the question. The test will have four short-answer questions that count towards your final score. Each question will be worth **2 points**.

EXTENDED-RESPONSE QUESTIONS
An extended-response question requires a longer response. Usually, an extended-response question has more parts to answer than a short-answer question. There will be two extended-response questions on the test that count towards your score. Each will be worth **4 points**.

CHAPTER 1

OHIO'S SCIENCE STANDARDS

The **Grade 8 Science Achievement Test** covers six science standards that you have studied in the sixth, seventh, and eighth grades:

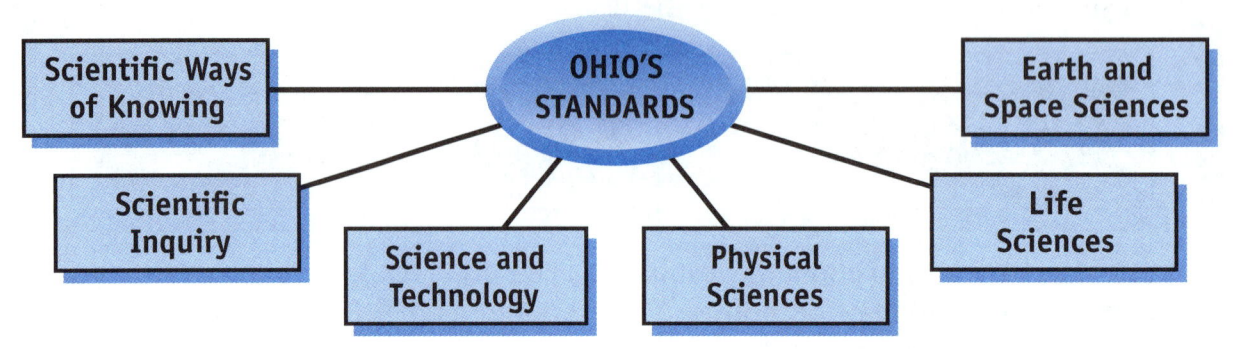

- Scientific Ways of Knowing
- Scientific Inquiry
- Science and Technology
- Physical Sciences
- Earth and Space Sciences
- Life Sciences

(OHIO'S STANDARDS)

Each standard has several "benchmarks" you should know. The questions on the test will focus on these benchmarks. The chart below shows you how many questions will be asked about each standard on the test. It does not include the four additional practice questions that will not count towards your final score.

Standard	Multiple-Choice Questions (1 point each)	Short-Answer Questions (2 points each)	Extended-Response Questions (4 points each)	Total Points
Scientific Ways of Knowing; Scientific Inquiry; Science and Technology	4 to 6	1 or 2	1*	10 to 14
Physical Sciences	8 to 10			10 to 14
Life Sciences	8 to 10	2 or 3	1*	10 to 14
Earth and Space Sciences	8 to 10			10 to 14
Total Number of Items (38)	**32**	**4**	**2**	**38**

On any test, only 1 of these 3 standards will have an extended-response item.

As you can see, the **Ohio Grade 8 Science Achievement Test** is a balanced test. It tests all three of the main fields of science equally. It also gives the same weight to the three standards about the general nature of science.

CHAPTER 1: OHIO'S SCIENCE STANDARDS   3

HOW THIS BOOK CAN HELP YOU

This book provides a complete "refresher" of the knowledge and skills you will need to perform your best on the Grade 8 Science Achievement Test.

TOOLS FOR MASTERING THE TEST

In Chapter 2, you will learn about the four types of questions on the test. Chapter 3 will show you how to answer short-answer and extended-response questions.

A REVIEW OF THE BENCHMARKS

The main part of the book consists of four units. Each unit covers a different Ohio science standard.

UNIT 2: THE NATURE OF SCIENCE

This unit consists of several chapters dealing with *Scientific Ways of Knowing*, *Scientific Inquiry* and *Science and Technology*. In this unit, you will learn how scientists think and approach problems. You will also learn the difference between scientific description and explanation, the differences between various types of scientific investigations, how to interpret data, and how technology affects our quality of life.

Students work together on a class experiment while using good safety practices.

UNIT 3: PHYSICAL SCIENCES

In this unit, you will learn about matter and energy. You will learn about the different forms of matter, their physical properties, and the differences between physical and chemical changes. You will also learn how force causes changes in motion, and how energy takes different forms.

UNIT 4: LIFE SCIENCES

In this unit, you will learn about living organisms. You will learn that living things are made of cells. You will learn how cells carry out their functions, and how plant and animals cells differ. You will also learn about inherited traits and both sexual and asexual reproduction. Finally, you will learn how organisms interact and adapt in ecosystems.

UNIT 5: EARTH AND SPACE SCIENCES

This unit explores the solar system, including the planets, comets, and asteroids. You will also learn about distant galaxies and stars. Then you will study Earth's interior, atmosphere, oceans, and landforms, and how they interact. Finally you will learn about the processes that contribute to the continuous changing of Earth's surface.

THE ORGANIZATION OF THE BOOK

Each content unit of this book has the following features:

★ Each chapter opens with a list of *Major Ideas*, highlighting the most important information in the chapter.

★ *Applying What You Have Learned* activities help you to apply what you have just learned in the chapter.

★ Each chapter finishes with a *What You Should Know* box, summarizing its major ideas.

★ Each chapter includes several *Study Cards* that highlight and help you review the most important *facts*, *concepts*, and *relationships* in the chapter.

★ Each chapter provides a *Checking Your Understanding* with OAT test-style practice questions. Each question is identified by its *standard*, *benchmark*, and *grade-level indicator*. The answer to the first question is fully explained for guided practice. Other questions provide independent practice.

★ Each unit includes a *Concept Map*, which visually summarizes the most important information for that standard.

★ Each unit concludes with a *Checklist* of relevant benchmarks, identifying all the information students need to know on that topic from sixth, seventh and eighth grade science.

FINAL PRACTICE TEST

The last part of the book consists of a complete practice test in science, just like the actual **Ohio Grade 8 Science Achievement Test**.

CHAPTER 2

HOW TO ANSWER MULTIPLE-CHOICE QUESTIONS

There will be four types of multiple-choice questions on the Grade 8 Science Achievement Test. These questions will test your ability to:

- Recall scientific facts, concepts and relationships
- Understand and analyze scientific information
- Demonstrate investigative processes of science
- Apply concepts and make connections with science

This chapter will help you to recognize each type of question. You will also learn a method to help you answer each type of question on the test. Let's begin by looking at the first type of question.

RECALLING SCIENTIFIC INFORMATION

Many questions on the test will simply ask you to *recall* or *identify* scientific facts, concepts and relationships. For example, examine the question below:

1. There are many different objects in our solar system, including planets, moons, asteroids, and comets. How do comets differ from these other objects?
    A. They orbit the sun.
    B. They are balls of superheated gases.
    C. Their orbits are determined by gravitational forces.
    D. Some of their ice turns to gas when they approach the sun.

As you can see, this question tests your ability to recall information about comets.

UNLOCKING THE ANSWER

- What do you think is the answer to Question 1? _____
- Explain why you selected that answer. _____
  _____
  _____

HOW TO USE THE "E-R-A" APPROACH

Whatever type of question you are asked, we suggest you follow the same three-step approach to answer it. Think of this as the "**E-R-A**" approach:

- **E**XAMINE the question
- **R**ECALL what you know
- **A**PPLY what you know

Let's take a closer look at each of these steps to see how they can help you to select the right answer.

STEP 1: EXAMINE THE QUESTION

Start by carefully reading the question. Be sure you understand any information the question provides. Then make sure you understand what the question asks for.

> **HINT**: This question asks you to select a unique feature of comets — "*How do comets differ from these other objects?*"

STEP 2: RECALL WHAT YOU KNOW

Next, you need to identify the topic of science that the question asks about. Take a moment to think about what you know on that topic.

Mentally review the important concepts, facts, and relationships you can remember.

> *In this case, you should think about what you can recall from your study of space. The question asks how comets differ from other objects in our solar system.*
>
> ★ *You may remember that most of the objects in our solar system orbit the sun.*
> ★ *Comets are balls of dust and gas that orbit the sun in very large orbits.*
> ★ *You should also recall that when a comet gets closer to the sun in its orbit, some of its ice turns to gas, creating a long, bright tail that is visible from Earth.*

STEP 3: APPLY WHAT YOU KNOW

Finally, take what you can recall about the topic and apply your knowledge to answer the question. Sometimes it helps to try to answer the question on your own **before** you look at the four answer choices. Then look to see if any of the answer choices is what you think the correct answer should be.

Review all the answer choices to make sure you have identified the best one. Eliminate any answer choices that don't make sense or that are obviously wrong. Then select your final answer.

> **HINT** *Here the question asks you how comets differ from other objects in our solar system. To answer this question, you need to recall the characteristics of comets. Then you have to apply this information by selecting a characteristic of comets that does not apply to other objects.*
>
> *Start by eliminating any choice that says something true of **both** comets and other objects. Also eliminate any choice that says something that does **not** apply to comets. The following choices are therefore wrong:*
>
> ★ *Choice A is true of comets, asteroids and planets.*
> ★ *Choice B describes the sun and other stars, but not comets.*
> ★ *Choice C is also true of other objects in the solar system.*
>
> *Choice D is the correct answer. It describes a characteristic of comets that makes them different from other objects in our solar system.*

ANALYZING SCIENTIFIC INFORMATION

A second type of question found on the test will examine your ability to *analyze scientific information*. To *analyze* means to break something up into its various parts in order to understand it better. Very often, this kind of question will present you with scientific data or observations and ask you questions about them.

Analyzing *scientific information* questions may ask you to:

- Organize, summarize or evaluate information
- Make estimates
- Choose the best model to represent the results of an experiment
- Draw conclusions from information
- Describe patterns or relationships in observations or data

Let's look at a sample question asking you to analyze scientific information.

2. Friedrich Mohs developed a mineral hardness scale in 1812. A harder mineral is able to scratch a softer mineral. A diamond, with a hardness of 10, is the hardest mineral found in nature.

MOHS' SCALE OF MINERAL HARDNESS

Hardness Rating	Mineral	Hardness Rating	Mineral
1	Talc	6	Feldspar
2	Gypsum	7	Quartz
3	Calcite	8	Topaz
4	Fluorite	9	Corundum
5	Apatite	10	Diamond

A scientist is attempting to identify an unknown mineral. The scientist has determined that the unknown mineral can scratch fluorite, but cannot scratch quartz. Which of the minerals from the table might this be?

A. talc
B. topaz
C. apatite
D. corundum

CHAPTER 2: HOW TO ANSWER MULTIPLE-CHOICE QUESTIONS   9

UNLOCKING THE ANSWER

🔑 Which minerals in the table are harder than fluorite? _____
_____

🔑 Which minerals in the table are softer than quartz? _____
_____

🔑 What is the answer to Question 2? _____ Explain why you selected that
answer. _____
_____
_____

As you can see, this question requires you to know how to read and interpret information found in a table and to draw conclusions from that information.

How should you answer a question that asks you to analyze scientific information? Again, you should try the "**E-R-A**" approach. Let's see how this approach could be used to answer this question.

USING THE "E-R-A" APPROACH

◆ **Step 1: EXAMINE the Question**

Read the question carefully. Be sure to examine any data or observations it may include. Usually, data will be presented as a diagram, graph, or table. Then be sure you understand what the question is asking for.

In question 2, the data is presented in a table. The question asks you to determine what the unknown mineral might be.

◆ **Step 2: RECALL What You Know**

Next, take a moment to think about any important facts, concepts and relationships you can recall about the subject of the question.

Here, you should recognize the question asks about **mineral hardness**. *Think about what you can remember about this topic. A harder mineral is able to scratch a softer mineral. In this case, this information is provided to you in the explanation preceding the table.*

continued

USING THE "E-R-A" APPROACH

◆ **Step 3: APPLY What You Know**

Apply what you know to select the correct answer. You might first attempt to answer the question without looking at the answer choices. Then examine the answer choices and eliminate those you know are wrong. Finally, select the best answer from the remaining answer choices.

In this case, you should be able to determine the answer from the information provided in the table. You should also be able to eliminate some of the wrong answer choices. In this question:

★ *Choice A is clearly wrong. Talc, with a hardness rating of 1, is too soft to scratch fluorite, with a hardness rating of 4.*

★ *Choice B is also obviously wrong. Topaz is too hard. It can scratch quartz, which the unknown mineral cannot do.*

★ *Choice D is also wrong. Corundum is too hard. It too can scratch quartz. Again, recall that the unknown mineral cannot scratch quartz.*

★ *Choice C is the correct answer. The scientist knows that the unknown mineral can scratch fluorite. Next, we know that the unknown mineral is unable to scratch quartz. These clues tell us that the unknown mineral must be harder than fluorite but softer than quartz. The only mineral among the four choices that meets these two criteria is apatite.*

EXPLAINING SCIENTIFIC INVESTIGATION

Some questions on the **Grade 8 Science Achievement Test** will examine your ability to think like a scientist. Here are some of the things you may be asked to do:

- Make observations
- Describe the procedures used in an experiment
- Make scientific measurements
- Make predictions or develop questions based on a scientific experiment

For example, look at the following question about a scientific investigation:

CHAPTER 2: HOW TO ANSWER MULTIPLE-CHOICE QUESTIONS   11

3. The diagram below shows two germinating corn seeds that have been placed in identical bottles and kept in the dark. Bottle A will be given a quarter turn each day for the next six days. Bottle B will not be turned.

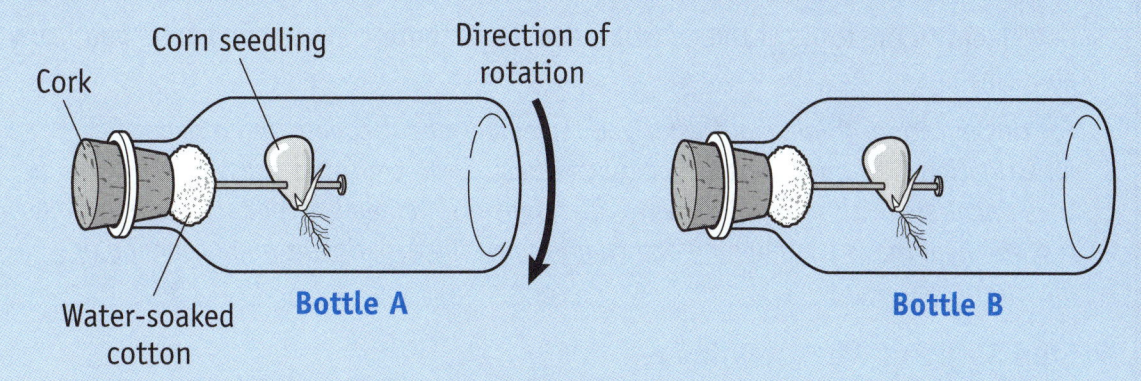

Bottle A — Cork, Corn seedling, Direction of rotation, Water-soaked cotton

Bottle B

Which hypothesis is being tested in this experiment?
A. Gravity has an effect on plant growth.
B. Water is needed for proper plant growth.
C. Enzymes aid in promoting seed development.
D. The amount of light received affects chlorophyl production.

UNLOCKING THE ANSWER

🔑 What is a hypothesis? _____

🔑 What do you think is the answer to Question 3? _____ Explain why you selected that answer. _____

USING THE "E-R-A" APPROACH

◆ **Step 1: EXAMINE the Question**

First, examine any information in the question. Then determine what the question is asking for.

In question 3, you are told the procedures followed in conducting an experiment. The question then asks you to determine what hypothesis is being tested in the experiment.

USING THE "E-R-A" APPROACH

◆ **Step 2: RECALL What You Know**

Now, identify the topic of the question and recall those concepts and facts you know about it.

For this question, think about what you remember about conducting a scientific investigation. You might recall that a hypothesis is an educated guess or explanation that scientists test in an experiment. You might also remember that scientists usually test the hypothesis by changing one condition in the experiment and observing or measuring the result.

◆ **Step 3: APPLY What You Know**

Finally, apply your knowledge to the question. Eliminate any wrong answer choices. Select the best answer choice remaining.

If you recall what a hypothesis is, then you can answer the question easily. Even if you don't remember what a hypothesis is, you may still be able to select the correct answer. Start by trying to eliminate those choices that you know are wrong. In the experiment, only one condition was changed: one bottle was partially turned each day for six days. This would change the direction the seedling is growing. There is no light in the room or enzymes, so **Choices C** *and* **D** *are incorrect. Each bottle has the same water, making* **Choice B** *incorrect. Since none of these are correct, the best answer must be* **Choice A**.

APPLYING SCIENTIFIC CONCEPTS TO "REAL WORLD" SITUATIONS

Some questions on the test will ask you to apply scientific concepts to "real-world" situations. Questions dealing with "real world" situations usually ask you to:

- Apply your scientific knowledge to new situations
- Use scientific concepts to solve problems
- Determine which scientific procedures to use in an investigation

Let's look at a sample question that asks you to make connections between science and the "real world."

CHAPTER 2: HOW TO ANSWER MULTIPLE-CHOICE QUESTIONS   13

4. In many parts of the world, farmers often face erosion from sudden, heavy rains. What would be the best way for farmers to plow their fields on a hill to reduce soil erosion?

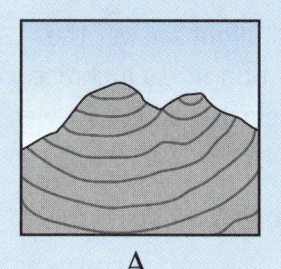

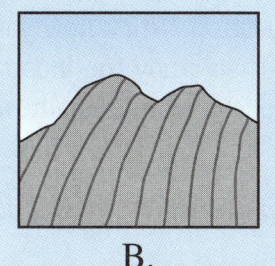

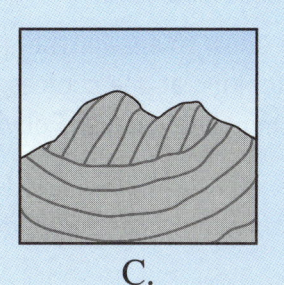

A.   B.   C.   D.

UNLOCKING THE ANSWER

🔑 What do you think is the answer to Question 4? _____

🔑 Explain why you selected that answer. _____
_____

To answer questions asking you to apply scientific concepts to solve a problem, you should again try using the "**E-R-A**" approach:

USING THE "E-R-A" APPROACH

◆ Step 1: EXAMINE the Question

Examine the question carefully. Be sure that you understand the "real world" situation presented in the question.

In this question, how can farmers plow their fields on a hill to keep soil erosion from heavy rainfall at a minimum?

◆ Step 2: RECALL What You Know

This kind of question asks you to apply your scientific knowledge to a "real-world" situation. You need to identify which scientific facts, concepts, and relationships best apply to the situation in the question.

To answer this question, you need to recall what you know about the topic of soil erosion. You might recall that rain works with gravity to carry soil downward off the sides of hills.

continued

USING THE "E-R-A" APPROACH

◆ **Step 3: APPLY What You Know**

Now use what you know to answer the question. Think how you might answer the question without looking at the choices. Then study the illustration in each choice carefully, eliminating those that are obviously wrong. Finally, select the best answer.

In this example:

★ **Choice B** would result in the soil being washed downward along the ridges created by plowing.

★ **Choices C** and **D** would generally have the same effect, since eventually the ridges lead downward.

★ **Choice A** is the best answer. By plowing along horizontal ridges (known as contour plowing), farmers help the soil resist the downward movement of the water.

Now use the "E-R-A" approach to answer a question on your own.

5. Several scientists wish to check the accuracy of experimental results published in another country. How could they check this data for accuracy?
   A. make a model to explain the data
   B. draw their own conclusions from the data
   C. repeat the experiment using the same procedures
   D. try a different experiment to test a new hypothesis

APPLYING WHAT YOU HAVE LEARNED

Use the "E-R-A." approach to help you answer question 5.

◆ **Examine the Question:** What does it ask? _____

◆ **Recall What You Know:** What do you remember about the topic? _____

◆ **Apply What You Know:** Which is the best answer choice? ☐

In this chapter, you learned how to answer different types of multiple-choice questions. In the next chapter, you will learn how to answer short-answer and extended-response questions.

CHAPTER 3

HOW TO ANSWER SHORT-ANSWER AND EXTENDED-RESPONSE QUESTIONS

Some questions on the **Ohio Grade 8 Science Achievement Test** will ask you to write out the answer in your own words.

SHORT-ANSWER QUESTIONS

There will be four short-answer questions on the test. A short-answer question requires you to give two facts or pieces of information. Each question is worth two points. These short-answer questions will test the following standards:

One or two short-answer questions will test the following:
- Scientific Inquiry
- Scientific Ways of Knowing
- Science and Technology

Two or three short-answer questions will test the following:
- Earth and Space Sciences
- Life Sciences
- Physical Sciences

EXTENDED-RESPONSE QUESTIONS

There will also be two extended-response questions on the test. An extended-response question will require you to give four facts or pieces of information. Each of these questions will be worth four points.

One extended-response question will test one of the following:
- Scientific Inquiry
- Scientific Ways of Knowing
- Science and Technology

A second extended-response question will test one of the following:
- Earth and Space Sciences
- Life Sciences
- Physical Sciences

RESPONDING TO A SHORT-ANSWER OR EXTENDED-RESPONSE QUESTION

There are many ways to approach short-answer and extended-response questions. One of the best ways is to use three main steps:

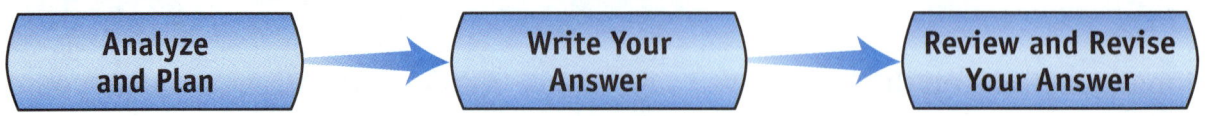

Analyze and Plan → **Write Your Answer** → **Review and Revise Your Answer**

STEP 1: ANALYZE AND PLAN

To answer either a short-answer or extended-response question, first look carefully at the directions of the question. The exact instructions for what you are supposed to do will usually be found in the "**action words**" of the question:

SOME OF THE MOST COMMON "ACTION WORDS"

Compare	To identify similarities and differences between two or more things.
Describe	To give the characteristics of something, tell what something is like, or tell how something changes over time.
Explain	To *explain how* something happened, *why it happened* or to *explain its effects*: • To *explain how*, tell the way in which it took place. • To *explain why*, give the reasons why it happened. • To *explain effects*, identify and describe each effect.
Identify	To name something, or to tell what it is.
Predict	To make a statement about what will probably happen in the future.
Support	To give facts or examples to back up a conclusion or point of view.
Draw	To create a diagram, map, or illustration of something.

After you have studied the "action words" of the question, you should next identify **all** of the parts of the question. Don't just rush into answering the question. Take a few moments to plan your answer. You can do this by simply jotting down a few notes you think might be helpful.

CHAPTER 3: HOW TO ANSWER SHORT-ANSWER AND EXTENDED-RESPONSE QUESTIONS

USING AN ANSWER BOX

It often helps to plan your answer with an **answer box**. This box divides the different parts of the question. You can fill in the answer box with your ideas or simply check off each part of the box as you answer it. The answer box serves as a checklist, making sure that you answer each part of the question. Even if you decide not to write out an answer box, you should do this process in your head:

★ **Short-answer questions** will have **two** parts. For example, a short-answer question might ask you to **describe** how *two types of rock are formed*. Here is what an answer box might look like for this question:

First Type of Rock	*Your response on how it is formed.*
Second Type of Rock	*Your response on how it is formed.*

★ **Extended-response questions** will usually have **four** parts. For example, an extended-response question may ask the following:

> Scientists continue to collect data that gives them a better understanding of Earth's interior and how Earth changes.
>
> In your **Answer Document**,
> - Describe two parts of Earth's interior.
> - Describe tectonic plates.
> - Identify one effect of tectonic plate movement. (4 points)

Here is what an answer box might look like for this question:

TASK		YOUR RESPONSE
Action Word	**What It Asks For**	**Fill in the boxes below:**
Describe	One part of Earth's interior (1 point)	
Describe	A second part of Earth's interior (1 point)	
Describe	Earth's tectonic plates (1 point)	
Identify	One effect of tectonic plate movement (1 point)	

18 MASTERING THE OHIO GRADE 8 SCIENCE ACHIEVEMENT TEST

Look at the following short-answer question. In the empty space below, create an answer box that could be used to answer the question. You **do not** need to fill in the boxes you create with specific information at this time: you will learn more about plant and animal cells later in this book.

> A biologist has discovered cells from an unknown living organism.
>
> In your **Answer Document**, describe two ways she can tell whether these are plant or animal cells. (2 points)

STEP 2: WRITE YOUR ANSWER

The next step in responding to a short-answer or extended-response question is to *write* your answer.

★ You can use the notes you created in your answer box to write your answer.

★ It may help to "echo" the question to introduce your answer. To **echo the question**, repeat it in the form of a positive statement. For example, suppose you were responding to the short-answer question on the top of this page. If you were to echo this question, your answer might begin:

> *There are two ways in which the biologist can tell whether these are plant or animal cells.*

★ Finally, turn each point in your notes or answer box into one or more complete sentences. Check off sections of your answer box each time you complete that part of your answer.

CHAPTER 3: HOW TO ANSWER SHORT-ANSWER AND EXTENDED-RESPONSE QUESTIONS

STEP 3: REVIEW AND REVISE YOUR ANSWER

The first person to read your answer should be **YOU** — *not* the person scoring it. Once you finish, read over your answer *before* you hand it in. Be sure you provide all of the needed information. As you review your answer, ask yourself:

★ Did I follow **all of the directions** in the question?

★ Did I complete **all of the parts** of the question?

★ Did I **provide enough details, examples**, and **reasons** to support my answer?

HOW YOUR ANSWER WILL BE SCORED

To see how your answer will be scored, let's look at a model question based on Benchmark E of Earth and Space Sciences:

> Scientists continue to collect data that gives them a better understanding of Earth's interior and how Earth changes.
>
> In your **Answer Document**,
>
> - Describe two parts of Earth's interior.
> - Describe tectonic plates
> - Identify one effect of tectonic plate movement. (4 points)

Read each of the following answers carefully. Then give each one a score of **0, 1, 2, 3, or 4** — with "4" as the best score. In this example, students should receive one point for each part of the question they answer correctly (*see the Answer Box on page 17*).

RESPONSE A:

> Earth has two parts: land and water. The land part includes mountains, valleys and deserts. Earth's water includes oceans, rivers and lakes. Most of Earth's life forms are found in the water. Water covers more of Earth's surface than the land.
>
> Tectonic plates are large slabs of rock that slide back and forth on Earth's surface. Their movements have created the continents.

Your score: ☐   Explain why you gave that score. _____

_____

RESPONSE B:

> Scientists have collected data that gives them a better understanding of Earth's interior. They call the center of Earth the core. The core is made of super-hot iron and nickel. Although the core would be hot enough to boil at the surface, pressure keeps the inner core solid and the outer core liquid. Around the core is an area of hot, semi-solid rock known as the mantle.
> Tectonic plates are pieces of Earth's crust that slip and slide over the mantle beneath. The movements of these tectonic plates have many effects on our planet. One of these effects is earthquakes.

Your score: ☐   Explain why you gave that score. _____

_____

RESPONSE C:

> Tectonic plates: slabs of rock
> Effect: helps to create mountains

Your score: ☐   Explain why you gave that score. _____

_____

RESPONSE D:

> A. Two parts of Earth's interior:
>   Crust. The outermost part of Earth. A thick skin of 5 to 50 km thick.
>   Core. The center of Earth. Made up of iron ore and other metals.
> B. Tectonic Plates: Thick slabs of solid rock that slowly float back and forth on Earth's surface.
> C. Effects of Tectonic Plate Movement: Folding of Earth's crust at convergent boundaries.

Your score: ☐   Explain why you gave that score. _____

_____

CHAPTER 3: HOW TO ANSWER SHORT-ANSWER AND EXTENDED-RESPONSE QUESTIONS   21

RESPONSE E:

> Earth's interior includes the crust. This is the thick skin of rock that covers Earth's surface. Tectonic plates are part of Earth's interior. They have many effects.

Your score: ☐   Explain why you gave that score. _____
_____
_____

Test scorers will use a **rubric**, or scoring guide, to score student responses on the Grade 8 Science Achievement Test. The rubric tells a scorer what information an extended-response should include to receive a score of **0, 1, 2, 3,** or **4 points**. Look at the following rubric for the question you just scored:

STANDARD AND BENCHMARK ASSESSED

Standard: Earth and Space Sciences

Benchmark: Explain the processes that move and shape Earth's surface.

Rationale: This question asks students to identify parts of Earth's interior and to relate tectonic plate movement to changes on Earth's surface.

The response will receive full credit if it describes two parts of Earth's interior, describes tectonic plates, and identifies one effect of tectonic plate movement.

STUDENT ANSWERS MAY INCLUDE:

Earth's Interior

★ **Crust.** This is the outer skin of Earth. It is made up of solid rock and is from 5 to 50 km deep.

★ **Mantle.** This is the hot, semi-solid rock that is found between the crust and Earth's core.

★ **Core.** This is the center of Earth. It is extremely hot. The inner core of Earth is solid, mainly iron, while the outer core is liquid iron and nickel.

Tectonic Plates. These are thick slabs of lithosphere, about 100 km thick, that slide back and forth over the mantle.

Some Effects of Tectonic Plate Movement

★ Earthquakes
★ Shape of continents
★ Rift valleys
★ Folding of crust to make mountains
★ Volcanoes
★ Mid-Atlantic Ridge

✓ Scoring Guidelines

Points	Student Response
4	The response describes two parts of Earth's interior AND describes tectonic plates AND identifies one effect of their movement.
3	The response describes two parts of Earth's interior AND describes tectonic plates OR identifies one effect of their movement. OR The response identifies one part of Earth's interior AND describes tectonic plates AND identifies one effect of their movement.
2	The response describes two parts of Earth's interior. OR The response describes one part of Earth's interior AND describes tectonic plates or an effect of their movement. OR The response identifies tectonic plates and an effect of their movement.
1	The response describes one part of Earth's interior. OR The response describes tectonic plates. OR The response identifies one effect of tectonic plate movement.
0	The response fails to demonstrate any understanding of Earth's interior or plate tectonics. The response does not meet the criteria required to earn 1 point. The response indicates inadequate or no understanding of the task and/or the information needed to answer the question.

Now that you have seen the rubric for scoring responses to this question, would you change any of the scores you gave before?

Which scores would you change and why? _____

CHAPTER 3: HOW TO ANSWER SHORT-ANSWER AND EXTENDED-RESPONSE QUESTIONS

Based on the rubric on pages 21 and 22, the responses would likely be scored as follows:

★ **Response A.** The student attempts to describe two parts of Earth's interior but in fact identifies two parts of Earth's surface. The first part of the response therefore should receive no credit. However, the student then goes on to identify tectonic plates correctly and to identify one effect of their movement. Since two of the four required points are provided in this response, it should receive a score of **2**.

★ **Response B.** The student successfully echoes the question to introduce the response. The response correctly describes two parts of Earth's interior. It also correctly describes tectonic plates and identifies one effect of their movement. Since all four of the required points are provided in this response, it should receive a score of **4**.

★ **Response C.** The student has written a very short response. Although it includes only ten words, it contains some of the required information. The student fails to adequately describe tectonic plates but the student does correctly identify an effect of tectonic plate movement. Since one of the four required points is provided, it should receive a score of **1**.

★ **Response D.** The student writes in outline form but provides all of the information needed to fully answer the question. The response correctly describes two parts of Earth's interior, describes tectonic plates, and identifies an effect of their movement. Since the response provides all four of the required points, it should receive a score of **4**.

★ **Response E.** The student correctly identifies one part of Earth's interior. However, the student does not describe tectonic plates adequately or identify any effects of their movement. Since only one of the required points is provided, the response should receive a score of **1**.

As you can see from this scoring exercise, the most important part of answering any short-answer or extended-response question is:

❋ reading the question carefully
        and
❋ answering **all the parts** of the question with **correct information**.

Notice also that the length of your answer does not determine the score you receive. The number of lines you write is far less important than providing the correct information that fully answers the question.

THE NATURE OF SCIENCE

UNIT 2

The **Ohio Grade 8 Science Achievement Test** will examine your knowledge of the general nature of science. To try to understand the natural world, scientists make observations, ask questions, form and test hypotheses, share ideas, and develop theories.

In the next three chapters, you will learn that scientific knowledge is not a fixed set of facts. In fact, it changes as scientists learn more about the world. You will also learn how scientists conduct their investigations, and how the results of these investigations are analyzed and communicated to others. Finally, you will learn how scientific knowledge and technology are closely related.

The process of scientific investigation.

CHAPTER 4: SCIENTIFIC WAYS OF KNOWING
In this chapter, you will learn about the nature of scientific knowledge. It is logical, based on evidence, and subject to change. Scientific knowledge can also be used to make predictions about the natural world.

CHAPTER 5: SCIENTIFIC INQUIRY
This chapter looks at the process of scientific investigation. You will learn how scientists form hypotheses and design different types of investigations to test their ideas. You will also learn about the need for laboratory safety, and how data from investigations is analyzed and interpreted.

CHAPTER 6: SCIENCE AND TECHNOLOGY
In this chapter, you will learn how science and technology are closely related. You will study how technology has advanced over time, and how technological solutions are designed.

CHAPTER 4

SCIENTIFIC WAYS OF KNOWING

In this chapter, you will learn what science is. You will also learn how science improves our understanding of the natural world.

— MAJOR IDEAS —

A. **Science** is a special way of investigating and explaining the natural world. Scientific knowledge is logical. It is based on evidence and can often be used to predict future events.

B. Scientists **observe nature** and conduct **investigations** to produce scientific evidence. They make **observations**, collect **data**, and carefully **record** their observations and measurements. Scientists use these methods to test their **hypotheses** and **theories**.

C. Scientists keep accurate records and **repeat** their investigations or have others repeat them to reduce the risk of **bias**.

D. Scientific **description** tells what scientists have observed or measured; scientific **explanation** tells how or why something occurred.

E. Men and women from all countries and cultures have contributed to the growth of scientific knowledge.

F. Scientific thinking is based on reasoning, creativity, skepticism, and openness. These qualities of scientific thinking are helpful in daily life.

WHAT IS SCIENCE?

Science is a special way of investigating and explaining the natural world:

★ Science is based on factual evidence provided by observations and data.
★ Science tries to find logical explanations for events in the natural world.
★ Scientific knowledge can be used to make predictions. Scientific ideas can often be tested by the accuracy of their predictions.

25

DESCRIPTION AND EXPLANATION

Accurate observation and description form essential aspects of science.

THE ROLE OF DESCRIPTION IN SCIENCE

Scientists describe the natural world by **making observations, taking measurements,** and **collecting data**. Sometimes they observe nature by simply watching what happens. For example, scientists may observe how tadpoles in a pond grow into frogs.

Scientists often **take measurements** so that they have an exact record of something. For example, they may measure the length, height, mass, hardness or temperature of an object so that they can later see how it changes. Often, scientists will **summarize** their observations or other data in order to describe something.

Scientific description consists of what scientists have actually observed or measured.

APPLYING WHAT YOU HAVE LEARNED

◆ What are some of the ways in which scientists can describe a natural event?
   _____
   _____

THE ROLE OF EXPLANATION IN SCIENCE

Scientists also try to **explain** why events happen in nature. They try to uncover underlying patterns or the reasons why natural events happen in the way that they do. Explanations are different from descriptions. A **description** tells what scientists have actually observed. An **explanation** is actually a type of **inference** or logical guess. An explanation connects together those facts and observations the scientists decide are most important. It tells why they happen. An explanation can also provide the basis for a **prediction** — what will happen in the future.

CHAPTER 4: SCIENTIFIC WAYS OF KNOWING

HOW EXPLANATION DIFFERS FROM OBSERVATION

For example, a scientist sits under an apple tree. He sees an apple fall to the ground. This is an observation. The scientist might describe the event very precisely. The scientist could measure the mass of the apple, the distance it fell, and the precise speed with which the apple fell to the ground. To explain **why** the apple fell, the scientist would have to go beyond observation and description. The scientist would have to make clear the reasons **why** the apple fell.

For many centuries, people did not really know why things fell. In the 1600s, **Sir Isaac Newton** came up with an **explanation**. He said all things are pulled towards Earth's center by the force of gravity. Unlike the falling apple, gravity is something we cannot actually see. Newton **inferred**, or concluded, that gravity exists because of all the falling objects around us. He argued that the same force also caused the orbits of Earth and other planets around the sun, and of the moon around Earth. Newton also used his explanation to make **predictions**. Based on his "law of gravity," he was able to predict where a fired cannon ball would fall to Earth.

Sir Isaac Newton

APPLYING WHAT YOU HAVE LEARNED

◆ Think over an experiment you have done in science class.
- **Describe** what you observed in the experiment. _____
  _____
- **Explain** what you saw in the experiment. _____
  _____

◆ On a separate sheet of paper, compare the processes of *observation* and *description* with *inference* and *explanation*.

THEORIES

A "big idea" in science is called a theory. A **theory** attempts to explain a large amount of data or a great number of observations. Newton's belief in the existence of an unseen force known as gravity is an example of an important theory. People sometimes refer to this idea as "Newton's Theory of Gravity."

A THEORETICAL VIEW OF THE UNIVERSE

Scientific theories affect how we understand the natural world. For example, Aristotle once taught that the sun and planets circle Earth. This theory, however, could not explain all the observations people made of the skies at night.

In the 1500s, a Polish astronomer, **Nicolaus Copernicus**, came up with a better idea. He argued that Earth and other planets moved around the sun. This theory explained why planets observed from Earth sometimes seemed to move backwards. Today, from the observations of telescopes and photographs taken in outer space, we can see that Copernicus was correct: Earth moves around the sun.

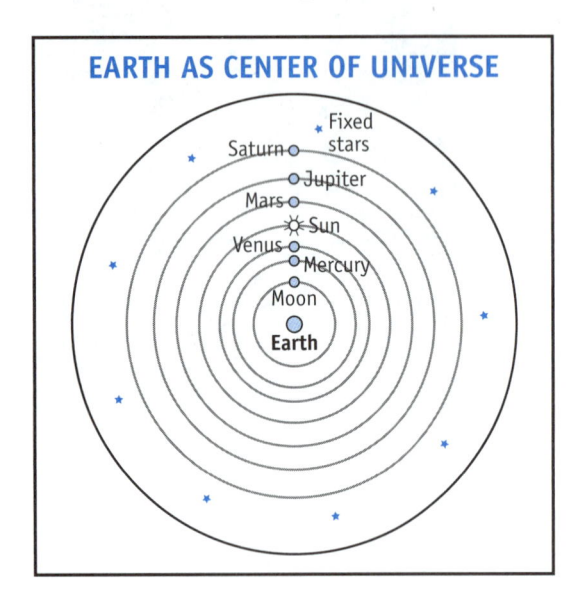

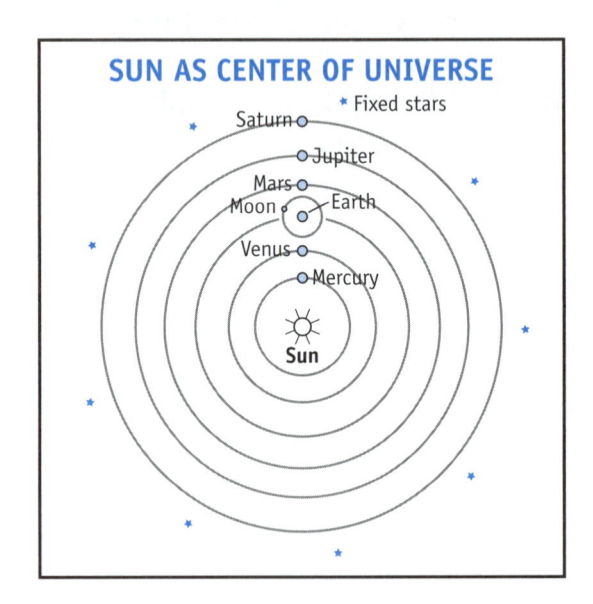

EARTH AS CENTER OF UNIVERSE

SUN AS CENTER OF UNIVERSE

THEORIES GUIDE RESEARCH

Theories guide research. For example, the French scientist **Louis Pasteur** found bacteria in spoiled food. He developed the theory that these bacteria caused the food to spoil. From this theory, he made a **hypothesis**, or educated guess. Pasteur predicted that if he applied heat to food, it would kill the bacteria and keep the food from spoiling. Pasteur found that by heating milk for several minutes, harmful microorganisms were killed. His hypothesis turned out to be correct. Today, milk is generally heated, or *pasteurized*, to kill bacteria.

How a Theory is Tested. One way to test a theory is to perform **experiments** to see if the theory can explain the results. Often the scientist will make a prediction from the theory. The scientist then conducts an experiment to see if the prediction is true. A valid theory should be able to explain all the results that scientists observe. One example is not enough to prove a theory, but it may be enough to disprove a theory. Theories must survive repeated testing. Otherwise, the theory may need to be rejected or changed. What if Pasteur had found that food spoiled without bacteria present? Then his theory would have been wrong.

CHAPTER 4: SCIENTIFIC WAYS OF KNOWING   29

APPLYING WHAT YOU HAVE LEARNED

◆ Why must a theory be changed or rejected if experiment results are not what it predicted? _____

_____

HOW SCIENTIFIC KNOWLEDGE CHANGES

There are many reasons why scientific explanations often change. Scientific instruments are regularly improved and new instruments are sometimes invented. These instruments enable scientists to discover previously unknown facts about the world that increase our scientific knowledge. For example, scientists could not see microscopic organisms until the invention of the microscope. They could not see the moons of other planets until the invention of the telescope.

Science is Ever-Changing. The more that scientists learn and observe, the more they refine their explanations of how nature really works. Scientists continually make observations about the world and ask more and more questions based on their observations. As new data is collected, scientists come up with new ideas. They develop better theories and explanations to describe the natural world.

APPLYING WHAT YOU HAVE LEARNED

◆ Identify an example where improved observations helped lead to new scientific ideas. _____

_____

RECORD-KEEPING AND BIAS

When scientists make observations and collect data, they **record** their results in writing. They need to record their results in order to provide evidence to support their theories and explanations.

Often, scientists keep their information in a **log** or **notebook**. These records must accurately state what the scientists have done and the results they received. Other scientists need to be able to review what has been done. These other scientists can then repeat the experiment or field investigation to see if they reach the same results.

Reducing the Risk of Bias in an Experiment. The ability of other scientists to repeat an experiment reduces the risk of **bias** — the possibility that a scientist's personal feelings have influenced the results. A scientist may be trying to prove or disprove a theory. Without realizing it, the scientist's attitude may influence how the experiment is conducted or the results are interpreted. This risk is reduced when the experiment is repeated many times by different individuals.

APPLYING WHAT YOU HAVE LEARNED

Define each term below to start your own glossary of scientific terms. As you progress through this book, add new terms of your own.

GLOSSARY ON THE NATURE OF SCIENCE

Scientific Term	Definition
Observation:	
Description:	
Explanation:	
Theory:	
Record-keeping:	
Bias:	

PEOPLE FROM ALL COUNTRIES AND CULTURES HAVE CONTRIBUTED TO SCIENCE

The development of scientific knowledge is a common human effort not restricted to any one country or culture. Men and women from every country and culture have contributed to the world's scientific understanding.

DIVERSE PEOPLE CONTRIBUTE TO SCIENCE

Below are some of the diverse group of people who have contributed to science:

Scientist	Background
Ibn-al-Haytham (Alhazen) (968–1038)	Alhazen was an Iraqi-born scientist who explained how lenses, mirrors and the human eye worked. He is considered the "Father of Optics" for his work with lenses and mirrors.
Anders Celsius (1701–1744)	This Swedish astronomer was responsible for devising a temperature scale based on the freezing and boiling points of water.
Sophie Germain (1776–1831)	Germain applied mathematics to explain how elastic objects vibrated. She became the first active woman attendee of the French Academy of Science.
Michael Faraday (1791–1867)	This English physicist built the first electric motor in 1821. He went on to invent the electric generator, a device that converts mechanical energy into electricity.
Elisha Gray (1835–1901)	Gray was a Professor of Dynamic Electricity at Oberlin College in Ohio. He invented a telephone at the same time as Alexander Graham Bell, but lost the legal battle over the patent.
George W. Carver (1864–1943)	George Washington Carver was an African-American botanical researcher. He helped the South diversify its agriculture and developed many new uses for peanuts.
Lise Meitner (1878–1968)	This Austrian female physicist was the first to realize that it was possible to split an atomic nucleus (nuclear fission). An element is named after her.
Alexander Fleming (1881–1955)	This Scottish biologist was responsible for discovering penicillin. In 1928, Fleming found that this mold could safely kill many bacteria that cause infections in humans.
Barbara McClintock (1902–1992)	This American biologist showed how genes can turn on and off to cause the physical characteristics of living organisms. For her work, she won the Nobel Prize in 1983.
Charles R. Drew (1904–1950)	Drew was an African-American doctor who helped to improve blood storage and to develop the first large blood banks during World War II.
Rachel Carson (1907–1964)	Carson was a marine biologist. Her pioneering book, *Silent Spring* (1962), is credited with launching the modern-day environmental movement.
Dorothy Hodgkin (1910–1994)	This English biochemist won the Nobel Prize for her work investigating the crystal structures of Vitamin B12, penicillin, insulin and other molecules with x-rays.

This short list of famous scientists shows how men and women from many different cultures and backgrounds have made important contributions to science.

APPLYING WHAT YOU HAVE LEARNED

Select any *three* scientists with different backgrounds. Base your research on the Internet or the use of an encyclopedia. Some scientists you might consider include: Aristotle, Archimedes, Isaac Newton, Robert Boyle, Joseph Priestley, Antoine Lavoisier, John Dalton, Louis Pasteur, Marie Curie, Albert Einstein, Stephen Hawking, and Shirley Jackson. For each scientist you select, briefly describe his or her background and how he or she made a contribution to the field of science.

Scientist	Background	Contribution to Science
1.		
2.		
3.		

THINKING LIKE A SCIENTIST

The work of science requires a variety of human abilities and qualities that are also very helpful in daily life.

REASONING

Scientists are very logical. They apply human reason to explain what happens in nature. This logical approach can be very useful in many everyday settings. For example, suppose you wake up one morning and find water on the floor of your room. You might approach this problem like a scientist. You would ask the question:

Where did this water come from?

Then you would consider all the ways the water might have come there:

★ Was there a leaking water pipe?
★ Was there a leak in the roof?
★ Did rain water come through the window?
★ Did someone spill water on the floor of your room?

Next, you would carefully investigate each possibility and decide whether or not it could be eliminated. You could check if there was a leaky pipe. You could ask your parents if it rained last night while you were asleep. After making observations and collecting data, you may come to a logical conclusion: for example, that you left the window open and it rained. Or you may be unable to determine what brought the water into your room, but at least you eliminated some of the possibilities.

CREATIVITY

We often think of musical composers, writers and artists as creative. Scientists must be creative as well. They are **creative** when they ask questions and think of different ways a natural event might have occurred. Scientists are creative again when they design investigations to see if their explanations are valid. This quality of creativity is very important in everyday life. Whenever you have a goal, try to be creative in thinking of ways to achieve it.

SKEPTICISM

Skepticism is the habit of not believing everything people tell you. Scientists are often skeptical. They listen carefully to others but do not accept everything they hear. They conduct their own experiments and make their own observations. For example, Copernicus did not accept the belief that the sun circled Earth. He doubted this and came up with a theory that explained more of the existing data: Earth moves around the sun. Some degree of skepticism is very useful in life. Skeptical people do not just accept what others say: they judge things for themselves.

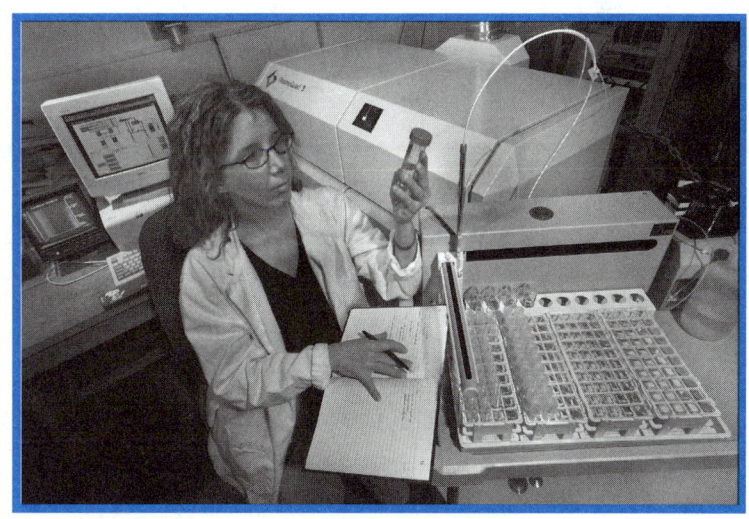

Scientists are skeptical and often need to make their own observations before accepting something.

OPENNESS

Scientists are very open. They do not keep their ideas or results secret from others. They criticize each other openly when they think errors have been made. This quality of **openness** helps science to advance. It is also a valuable quality for everyday life. People who are open share their ideas. They are also not afraid to speak up when a mistake has been made. This helps to ensure that the mistake is not repeated.

Because scientific thinking is useful in everyday life, the pursuit of scientific knowledge is also helpful for almost any career. A person with scientific knowledge not only understands the forces that control events in the natural world, but also understands how to approach the world scientifically — using logical reasoning, human creativity, skepticism, and openness to solve problems and reach goals.

APPLYING WHAT YOU HAVE LEARNED

◆ Explain how each of the qualities listed below is useful both in science and in everyday life.

Quality	How It Is Useful in Science and in Everyday Life
Reasoning	
Creativity	
Skepticism	
Openness	

◆ A number of people have fallen on the sidewalk in front of your aunt's house. Your aunt is very sorry, but she believes nothing can be done about this situation. How can you help solve this problem using the qualities of a scientist — logical reasoning, creativity, skepticism, and openness?

_____
_____
_____
_____

PROFILES IN SCIENTIFIC COURAGE

Scientists are not only logical, creative, skeptical and open-minded: they are also often courageous. It takes courage to stick by your own beliefs when others doubt you. It also takes courage to challenge accepted ideas. Throughout history, some scientists have displayed great courage in the face of persecution and personal danger.

GALILEO GALILEI (1564–1642)

Galileo, known as the "Father of Modern Science," was born in Italy during the Renaissance. Galileo began studying the movements of pendulums while still attending university. He questioned accepted ideas of motion and developed experiments that proved that heavier objects do not fall faster than lighter ones. Galileo was the first scientist to use a telescope to study objects in space. He published a book in support of Copernicus' theory. Galileo refused repeated orders from the Catholic Church to halt all discussions of his theories. He was put on trial for challenging accepted teachings of the Catholic Church and was sentenced to life imprisonment. He spent the last nine years of his life under house arrest.

XIE XIDE (1921–2000)

Xie Xide, a Chinese physicist, received her doctoral degree from MIT in the United States. After graduation, she returned to China. She made important discoveries in solid-state physics, semiconductors and compressed gases. In 1966, the Chinese Communist government took steps against scientists and other intellectuals. Xide, who supported cooperation between China and the United States, was suspected of disloyalty. Government leaders had her arrested and locked inside a cold laboratory for nine months. Later, she was forced to clean bathrooms and hallways at the university as a way to embarrass and humiliate her. When the campaign against intellectuals ended in the 1970s, Xide went back to teaching and founded China's Modern Physics Research Institute. She also helped establish Chinese-American educational exchanges and became President of China's Fudan University

WHAT YOU SHOULD KNOW

Place a (✔) next to those items you recall and understand. If you are unsure of an item, review the information in this chapter before moving on to the next chapter.

- [ ] You should know that **science** is a special way of investigating and explaining the natural world.
- [ ] You should know that scientists **observe nature** and conduct **investigations** to produce scientific evidence. They make **observations**, collect **data**, and **record** their observations in order to test their hypotheses.
- [ ] You should know that scientists keep accurate records and **repeat** their investigations or have others repeat them to reduce the risk of **bias**.
- [ ] You should know that scientific **description** tells what scientists have observed; scientific **explanation** tells how or why something occurred.
- [ ] You should know that men and women from all countries and cultures have contributed to the growth of scientific knowledge.
- [ ] You should know that scientific thinking is based on reasoning, creativity, skepticism, and openness. These qualities are helpful in daily life.

CHAPTER STUDY CARDS

Below are **Study Cards**. These are found at the end of each content chapter. You can copy or photocopy these cards. You are also encouraged to make your own cards to reinforce the most important terms and concepts in each content chapter.

Characteristics of Science

Science is a special way of investigating and explaining what happens in nature.

★ Science explains events in the natural world.

★ Science is based on evidence from making observations and collecting data.

★ Science can be used to help make predictions.

★ **Record-Keeping.** Scientists always keep accurate records so that others can check and repeat their investigations. This helps to reduce **bias** — the personal feelings a scientist may have that could influence the interpretation of results.

Scientific Description vs. Explanation

★ **Description.** Tells what scientists have observed and measured. Scientists often summarize their results to describe something. For example, a scientist sees an apple from an apple tree fall to the ground:

The apple fell a distance of one meter.

★ **Explanation.** Provides reasons **why** events happen by going beyond observation and description. Selects which information is most important. Explanations are **inferences** — conclusions drawn from the evidence. Can be used to make predictions. For example:

Gravity caused the apple to fall from the tree.

Scientific Knowledge

★ **Theories.** These are attempted explanations of many observations and large amounts of data.

★ Theories can be tested by further investigation and experimentation.

★ One example cannot prove a theory, but it can sometimes disprove a theory.

★ Scientific knowledge grows as technology improves scientific observations and as scientists test and revise their theories.

Characteristics of Scientists

Scientists have qualities that are useful not only in science but in everyday life:

★ Scientists use logical reasoning.

★ Scientists are creative and open.

★ Scientists are skeptical. They do not just accept what others tell them.

Diversity. Scientists come from a diversity of countries and cultures. Major contributions to science have been made by people from all races, religions and cultures.

CHECKING YOUR UNDERSTANDING

1. Which statement provides an explanation rather than a description?
   A. The water boiled because its molecules were moving faster.
   B. The temperature of the water was 100° Celsius when it boiled.
   C. The scientist used a stopwatch to measure how long the water took to boil away.
   D. The scientist heated a beaker with 500 mL of water until the water boiled.

> **HINT**
>
> This question asks you to tell the difference between a scientific description and an explanation. A scientific description includes observations, measurements, and summaries of this information. A scientific explanation tells **why** something happened. An explanation often includes an inference, prediction, or the selection of particular information as more important. In this example, **Choices B, C** and **D** describe particular events that seem to be part of the same experiment. A scientist has heated a beaker of water. The scientist then used a stopwatch to measure how long it took the water to boil away. Only one choice attempts to explain *why* the water boiled. The transfer of heat to the water made its molecules move faster and faster, until the water eventually boiled. Therefore, **Choice A** is the correct answer.

Now try answering some additional questions on your own.

2. A group of students is investigating what frogs eat. Each day the students observe a group of frogs in a local pond.

   Why is it important for the students to carefully record their observations?
   A. Frogs are in danger of becoming an endangered species.
   B. Readers will help create a more frog-friendly environment.
   C. Other scientists will know exactly what the students observed.
   D. The students will be able to obtain support from the government.

   SK: A
   G6.2

3. Which best explains why scientists need to keep accurate records of their work?
   A. It makes science appear more interesting.
   B. It helps organizations to keep track of all expenses.
   C. It allows other scientists to repeat their experiments.
   D. It is a government requirement for scientific experiments.

   SK: C
   G4.2

4. Which is an important characteristic of scientific thinking?
   A. secretive
   B. skeptical
   C. optimistic
   D. emotional

   ♦ Examine the Question
   ♦ Recall What You Know
   ♦ Apply What You Know

   SK: C
   G7.3

5. A group of students is conducting an experiment. At the end of their experiment, they obtain unexpected results. What should they do next?
   A. conduct a different kind of experiment
   B. ignore all of the results that were unexpected
   C. repeat the experiment to see if they reach similar results
   D. create a graph of their results and communicate it to others

   SK: B
   G7.2

6. What conclusion can best be reached from a study of the work of Michael Faraday, Elisha Gray, Rachel Carson, Dorothy Hodgkin, and Xie Xide?
   A. The greatest scientists have come from the United States.
   B. Men have made greater contributions to science than women.
   C. Most important scientific discoveries have already been made.
   D. People from many different cultures have contributed to science.

   SK: C
   G6.5

7. A scientist is conducting an experiment on the effects of a new virus on people. Which practice by the scientist shows the scientist is biased?
   A. The scientist studies the virus using a high-powered microscope.
   B. The scientist uses only research by scientists holding similar views.
   C. The scientist creates a statistical study of people exposed to the virus.
   D. The scientist injects humans with the virus after fully explaining all risks.

   SK: E
   G8.2

> A scientist gave a group of 10 chickens a special type of feed mixed with vitamins. Another group of 10 chickens received none of this special feed. When the scientist weighed both groups after a month, he found that the chickens that had received the special feed gained more weight than the others.

8. Which statement provides a description, rather than an explanation, of what took place?
   A. Chickens will eat more if they are hungry.
   B. Eating special feed can make some chickens grow plumper.
   C. Vitamins should be an important part of every chicken's diet.
   D. One group of chickens weighed more than the other after a month.

   SK: A
   G8.1

> A scientist is employed by a pharmaceutical company. The scientist discovers a unique drug that can help people fall asleep within five minutes. The scientist claims there are no negative side-effects to using the drug.

9. Why might other scientists be skeptical about such a claim?
   A. The scientist was not born in the United States.
   B. The scientist studied chemistry at a leading U.S. university.
   C. The scientist wanted more money from the pharmaceutical company.
   D. The scientist may be biased since he works for the company that will produce the drug.

   SK: B
   G8.2

10. Scientific thinking is not restricted to the laboratory. It is helpful to people in a variety of everyday settings.

    In your **Answer Document**,* describe two characteristics of scientific thinking that are helpful to people in daily life. (2 points)

    SK: C
    G7.3

11. Scientists keep accurate records when they are conducting an investigation.

    In your **Answer Document**,* provide two reasons why it is important that scientists keep clear, thorough and accurate records. (2 points)

    SK: A
    G6.2

12. Men and women from all countries and cultures have made important contributions to the development of science.

    In your **Answer Document**,* identify two individuals who have contributed to the development of science. For each individual you identify, describe one contribution that he or she made to the world's scientific understanding. (4 points).

    SK: C
    G6.5

   *Whenever you see a reference to "Answer Document" in this book, please write your response on a separate sheet of paper from your notebook.

CHAPTER 5

SCIENTIFIC INQUIRY

In this chapter, you will learn more about how scientific investigations are conducted. You will learn how scientists ask questions, develop and test hypotheses, analyze data, and draw conclusions.

— MAJOR IDEAS —

A. **Scientific inquiry** is the process by which scientists ask questions and investigate the natural world.

B. There is no single, fixed set of procedures for all scientific investigations. Each particular investigation will determine the procedures that are needed.

C. The following steps are often used by scientists to conduct an experiment:
    ★ The scientist observes the world and **asks a well-defined question**.
    ★ The scientist develops a **hypothesis** to answer the question.
    ★ The scientist designs an experiment to **test the hypothesis**. Often, an experiment changes one **variable** to see its effect on other variables. Sometimes, however, it is not possible to control all the variables.
    ★ The scientist safely uses appropriate **tools and technology** to conduct the experiment.
    ★ The scientist **organizes**, **analyzes**, and **interprets the results**. Scientists often use graphs, tables, and charts to see if there is a mathematical relationship between variables. Often there is more than one way the data can be explained.
    ★ The scientist **communicates** experimental procedures, results and conclusions to others.

D. Scientists take safety into account in all laboratory and field investigations.

METHODS OF SCIENTIFIC INVESTIGATION

Scientific inquiry begins when a scientist observes the natural world. The scientist then asks questions about what is observed.

For example, a scientist may see a burning candle and ask: "What happens when a candle burns?" Or the scientist may see a dog and her infant puppies and ask: "Why do children often resemble their parents?"

A VARIETY OF METHODS

As you learned in the last chapter, scientists are logical, creative, skeptical, and open-minded. They try to think of ways to design investigations to answer the questions they ask. There is, however, no single, fixed set of procedures that guides every type of scientific investigation. Scientists have made some of their greatest advances when they thought creatively and "broke the rules."

Scientists actually have a wide range of methods they can use to test an idea:

★ **Observe Nature.** Scientists can observe nature to see if they find examples that support their point of view.

★ **Conduct a Laboratory Experiment.** Scientists can conduct a laboratory experiment in which one variable is changed. Then they measure the effects this change has on a second variable.

★ **Conduct a Controlled Experiment.** Scientists can conduct a controlled experiment in which a variable is changed for one group but not the other.

★ **Conduct a Field Experiment.** Scientists can conduct field experiments in which they change a condition in a natural setting and then observe the effects this change has.

APPLYING WHAT YOU HAVE LEARNED

◆ Why do most scientists now agree that a single, fixed set of procedures cannot be applied to all scientific investigations? _____
_____
_____

THE ROLE OF MATHEMATICS

Many modern scientific theories are influenced by mathematics, which can be applied to anything that can be measured. Once scientists develop a model for how something works, they can apply mathematics to make predictions and to formulate hypotheses. Later in this chapter, you will learn how scientists use their mathematical skills to analyze data and draw conclusions.

STEPS OF A SCIENTIFIC EXPERIMENT

Scientists use many methods to test their ideas. The nature of an investigation determines the procedures that are needed. As you know, there is no single, fixed method for every scientific investigation. However, many scientists use the following steps to conduct an experiment:

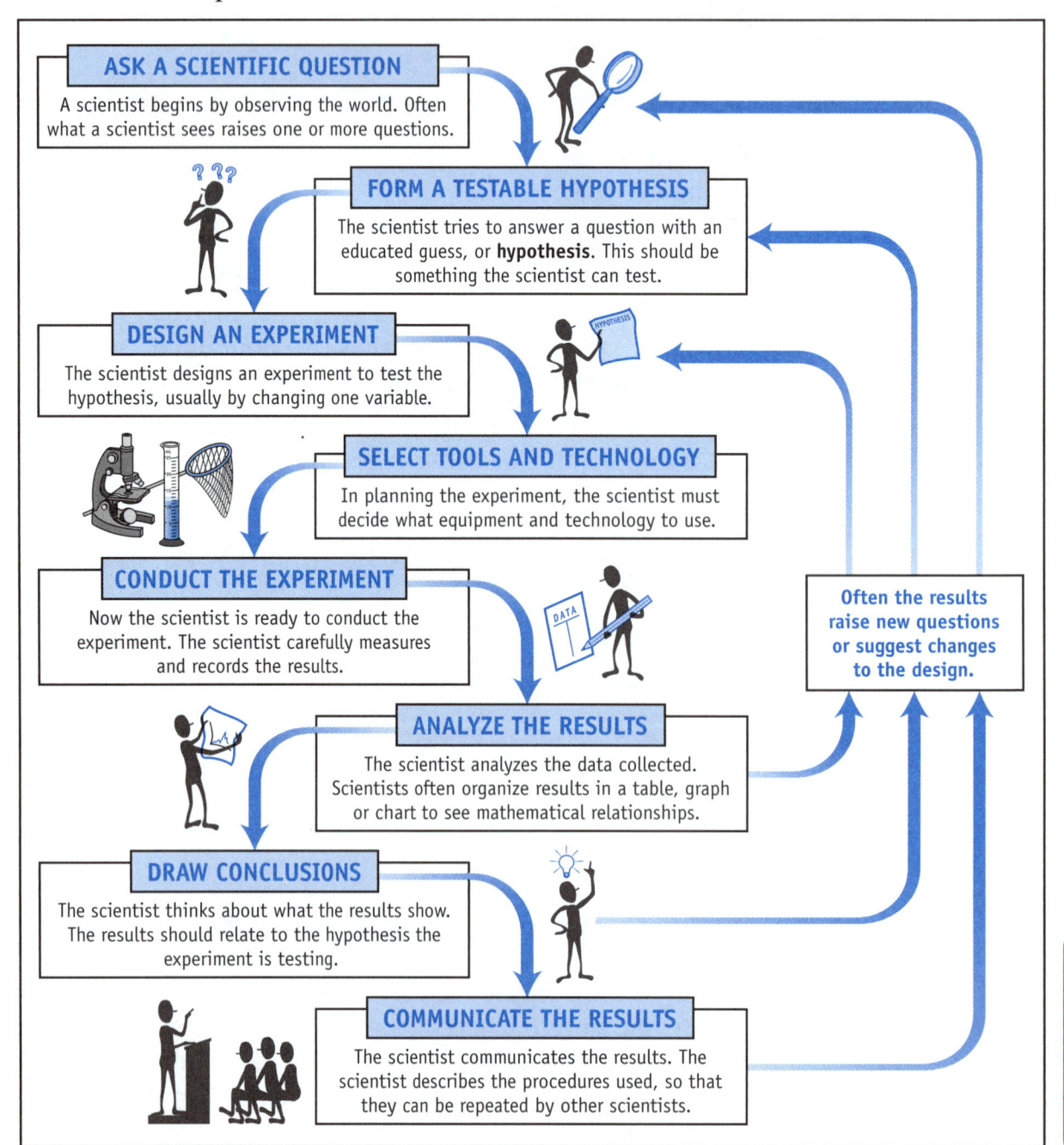

ASK A SCIENTIFIC QUESTION
A scientist begins by observing the world. Often what a scientist sees raises one or more questions.

FORM A TESTABLE HYPOTHESIS
The scientist tries to answer a question with an educated guess, or **hypothesis**. This should be something the scientist can test.

DESIGN AN EXPERIMENT
The scientist designs an experiment to test the hypothesis, usually by changing one variable.

SELECT TOOLS AND TECHNOLOGY
In planning the experiment, the scientist must decide what equipment and technology to use.

CONDUCT THE EXPERIMENT
Now the scientist is ready to conduct the experiment. The scientist carefully measures and records the results.

ANALYZE THE RESULTS
The scientist analyzes the data collected. Scientists often organize results in a table, graph or chart to see mathematical relationships.

DRAW CONCLUSIONS
The scientist thinks about what the results show. The results should relate to the hypothesis the experiment is testing.

COMMUNICATE THE RESULTS
The scientist communicates the results. The scientist describes the procedures used, so that they can be repeated by other scientists.

Often the results raise new questions or suggest changes to the design.

CHAPTER 5: SCIENTIFIC INQUIRY

The specific order of these steps may vary. Results from an experiment or investigation often cause scientists to form a new hypothesis. As they begin to collect data, scientists may learn new ways to improve their experimental design. Let's look at each of these steps in greater detail.

ASK SCIENTIFIC QUESTIONS

Questions that are vague or that ask for opinions cannot be answered by a scientific experiment. Questions for investigation must be *factual* and *specific*. When it is specific, a scientific question can often be answered through a scientific investigation. A good research question identifies *exactly* what will be identified in an experiment or field study. Equally important, good scientific questions relate back to key scientific concepts. The answer to the question should help support or disprove a scientific theory or explanation. Because they relate to scientific concepts, such questions help advance scientific knowledge.

FORM A TESTABLE HYPOTHESIS

As you learned in the last chapter, a **hypothesis** is an educated guess, often based on a theory. The hypothesis attempts to answer the question under investigation. A good hypothesis can be tested in an experiment. For example, a scientist may make the hypothesis that a certain type of chemical will prevent the growth of **bacteria** — microscopic organisms that can cause disease.

Often a scientist will use a theory to make a **prediction**. The prediction then serves as the hypothesis. For example:

If this chemical is placed with a sample of bacteria, the bacteria will die.

This prediction can be tested in an **experiment** — a test carried out under controlled conditions.

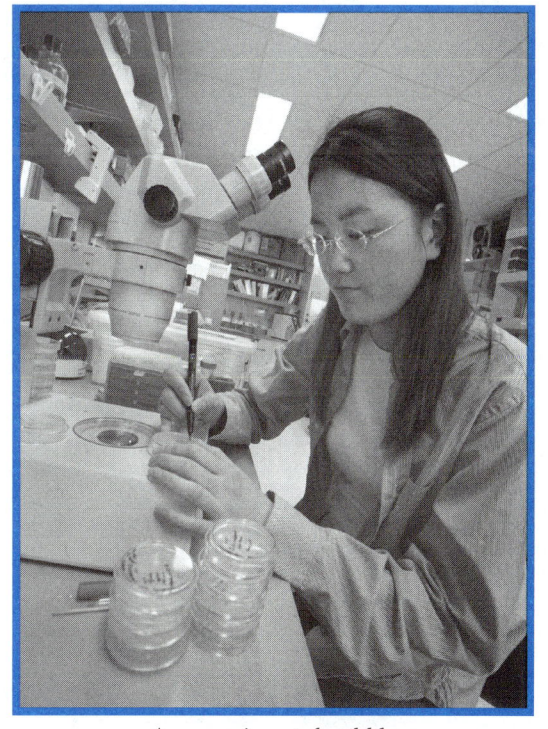

An experiment should be capable of testing the hypothesis.

An experiment may show that the original hypothesis is either right or wrong. In science, proving that a hypothesis is wrong can be just as valuable as proving that it is right.

APPLYING WHAT YOU HAVE LEARNED

◆ Think of an experiment you did in science class this year. What hypothesis did that experiment test? _____
_____

◆ Why can it be just as important to prove that a hypothesis is wrong as to prove that it is valid (*correct*)? _____
_____

DESIGN AN EXPERIMENT TO TEST YOUR HYPOTHESIS

An experiment creates **controlled conditions** to test the hypothesis. The **experimental design** is the roadmap of the experiment.

INDEPENDENT VARIABLES

The researcher must have a clear understanding of the variables in the experiment. A **variable** is anything that can **vary**, or change. Some variables can be changed by the experimenter. For example, a researcher testing whether a particular chemical kills bacteria can decide: what kind of bacteria to test, what temperature the bacteria samples should be kept at, and how much of the chemical to add to each sample. Each of these could be an example of an **independent variable**.

DEPENDENT VARIABLES

Other variables cannot be directly controlled by the experimenter. How much the bacteria grow is an example of this second kind of variable, known as the **dependent variable**. *In an experiment, the dependent variable changes in response to changes introduced to the independent variable.* Some scientists call this the "responding variable."

VARIABLES KEPT CONSTANT

In a controlled experiment, the experimenter usually changes only *one independent variable* at a time. For example, in an experiment to test the effect of a chemical on bacteria, a scientist will select one independent variable, such as whether or not the bacteria are exposed to the chemical. The researcher will change that variable to see how this change affects a second variable, such as the growth of bacteria in the sample.

Other variables in the experiment are kept the same. The researcher will use only one kind of bacteria in the samples. The experimenter will keep all the samples in identical petri dishes, at the same temperature, humidity and light. Everything will be kept the same as possible except for the tested variable.

EXPERIMENTS WITH CONTROL GROUPS

In many experiments, the scientist changes one variable to see the effect this has on another. In other experiments, the scientist keeps one sample the same and then introduces a change to one variable in a second sample. The unchanged sample is the **control group**. The changed sample is the **experimental group**. The scientist then compares the results of the experimental group with the control group.

Ideally, a scientist will change only one independent variable to see the effect this has on one dependent variable. In practice this is not always possible. For example, a manufacturer may be testing how well a new drug helps patients overcome a disease. Since every person is unique, it is impossible to bring all the variables under control. The manufacturer cannot find two individuals who are exactly the same in order to test what happens with or without the drug. To overcome this difficulty, scientists often use large samples in controlled experiments. If the manufacturer tests the new drug on thousands of people, individual differences will be likely to average out.

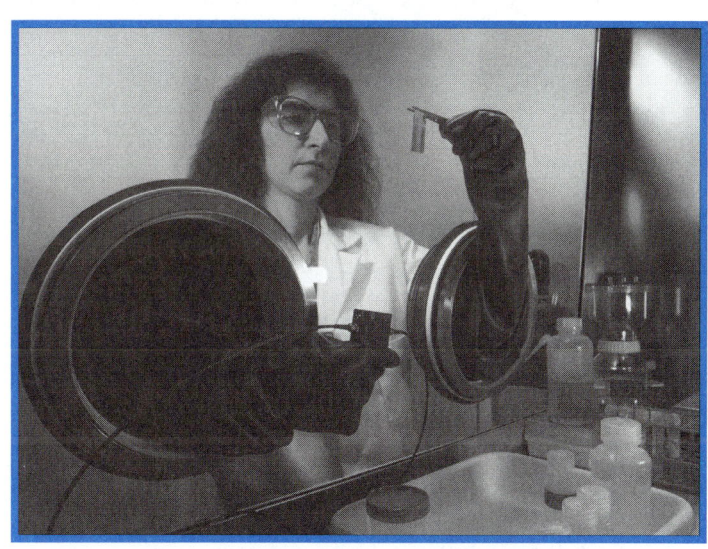

Manufacturers of new drugs must conduct tests on large samples of patients.

The *larger* the sample size, the *more accurate* the results. If we compare the reaction of just two people to a drug, the results may not accurately show the effect of the drug. If 1,000 patients taking the drug are cured and 1,000 patients in a control group without the drug remain ill, then it is very likely that the drug helped members of the experimental group to get better.

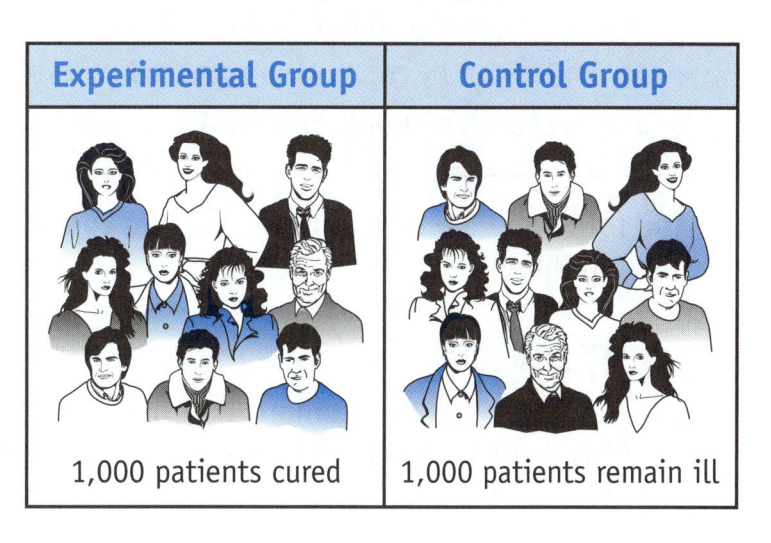

Experimental Group	Control Group
1,000 patients cured	1,000 patients remain ill

Ethical considerations may also influence testing. For example, if people are dying from a disease and early tests show that a new drug is successful in curing patients, doctors may suspend the test and give the people in the control group an opportunity to take the new drug.

APPLYING WHAT YOU HAVE LEARNED

- Why do scientists generally change only one independent variable in an experiment? _____
_____

- Why do some experiments have both an experimental and a control group?
_____

- How does sample size affect the validity of experimental results? _____
_____

SELECTING TOOLS AND TECHNOLOGY

Designing an experiment is a little like making a recipe. The experimenter must identify the required materials, including tools and technology. Then the experimenter must list the procedures to be performed to conduct the experiment.

ELEMENTS OF A GOOD EXPERIMENTAL DESIGN

★ All the variables need to be identified, and only one independent variable is experimental.
★ All the necessary materials are listed, including their amounts.
★ Variables can be precisely measured with instruments.
★ There should be numerous trials.

The following are some of the standard types of laboratory and field equipment you should know:

MAGNIFIERS

★ **Telescope.** An instrument that uses lenses and mirrors to magnify distant objects that appear small, such as planets or stars.

★ **Microscope.** A compound light microscope uses a series of lenses to magnify specimens placed on slides.

TOOLS FOR MEASURING

★ **Meter Stick.** A ruler usually marked in centimeters (cm), that is used to measure length.

★ **Graduated Cylinder.** A glass cylinder marked in milliliters (mL) to measure the volume of liquids.

★ **Balance.** An instrument with one or two pans used to measure the mass of an item in grams (g) or kilograms (kg).

★ **Spring Scale.** A scale that measures the weight of a hanging object that pulls down a steel spring.

★ **Thermometer.** An instrument used to measure temperatures in degrees Fahrenheit (F) or Celsius (C).

★ **Barometer.** An instrument that measures atmospheric pressure. Changes in barometric pressure usually indicate a change in the weather.

★ **Timer.** A clock, stopwatch or other device that precisely measures the passage of time.

SAFETY EQUIPMENT

★ **Safety Goggles.** Plastic goggles worn during an experiment to protect the eyes from dust, splashes, mists, or sprays.

★ **Laboratory Apron.** A bib worn over clothing to protect clothing and the skin from splashes, spilled chemicals or biological materials.

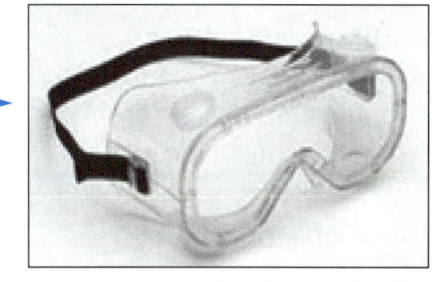

OTHER STANDARD EQUIPMENT

★ **Beaker.** A flat-bottomed glass container that is used in laboratories to hold liquids.

★ **Test Tube.** A small glass tube-shaped container rounded at one end and open at the other end. It is used to hold chemicals or heat substances in a laboratory.

★ **Petri Dish.** A circular, covered glass dish that is used to hold biological samples.

SAFETY PRECAUTIONS

Attention to safety is essential during both field and laboratory investigations. Safety must be considered even before an experiment begins, and sufficient safety precautions should be included in every experimental design. For example, before using any chemicals, scientists consult the proper **Material Safety Data Sheet** (MSDS) for information about each chemical. The MSDS identifies potential hazards and the appropriate protective equipment that should be used when handling each type of material. Important safety concerns include:

All scientists are quite concerned about following safety procedures in the laboratory.

★ How will certain chemicals be used and could they produce dangerous fumes?

★ Are there any biological hazards, and how should they be handled?

★ Should those who perform the experiment protect themselves with safety goggles and laboratory aprons? Is a fire extinguisher nearby?

Safety Equipment. Your classroom laboratory may have a fire extinguisher, fire blanket, first aid kit, safety goggles, and eye and face wash stations. You should know exactly where these are located and how they are used.

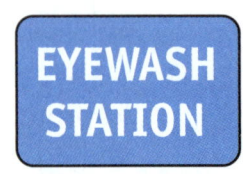

SOME COMMON LABORATORY SAFETY RULES

★ Follow your teacher's directions at all times.

★ Wear safety goggles, a laboratory apron, and gloves for many experiments.

★ Read all chemical labels and safety symbols.

★ Never heat liquids in closed containers. Always point the open end of test tubes away from others when heating liquids or mixing chemicals.

★ Wash hands with soap and water before and after all experiments, and know where the fire extinguisher is located.

CHAPTER 5: SCIENTIFIC INQUIRY   49

> ### APPLYING WHAT YOU HAVE LEARNED
>
> *Think about a laboratory experiment you conducted in school.*
>
> ◆ What hypothesis did you test? _____
> _____
>
> ◆ What laboratory equipment did you use? _____
> _____
>
> ◆ What safety precautions did you follow? _____
> _____

COLLECT DATA

Information collected during an investigation is known as **data**. Data can be in the form of precise measurements or simple observations. A scientist uses a notebook, diary or log during a laboratory experiment or field investigation to record results. When something is done, observed, or measured, an entry is made in the notebook. All the information entered is dated.

Taking Precise Measurements. The measurements recorded in the laboratory or field notebook come from using appropriate tools. Measurements are made using the international standards (*SI*) or **metric system**. For example, you would use a meter stick to measure the height of an object in centimeters. All measurements should be taken several times and recorded with their units (*for example, centimeters*).

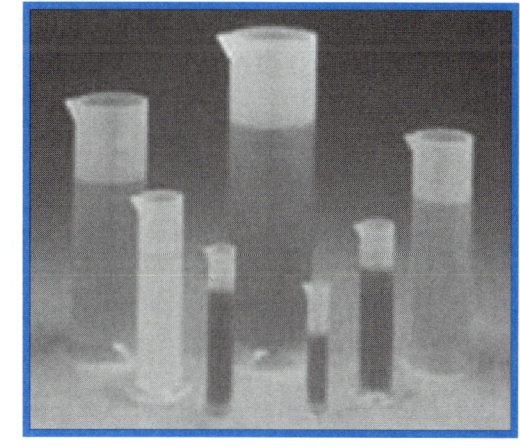

Several graduated cylinders

COMMON METRIC MEASUREMENTS

Measurement	Length		Mass		Volume	
Basic Unit	meter	m	gram	g	liter	L
Smaller Unit	centimeter	cm	milligram	mg	milliliter	mL
Same or Larger Units	kilometer	km	kilogram	kg	cubic centimeter	cm³

USING A BALANCE

A **balance** is an instrument used to measure mass. A **double-pan balance** has a pan on each side. The object to be measured is placed in one pan. Known units of mass are placed in the other. In this illustration, a mass of 500 gm and 100 gm are used. When the two pans are level, the masses in the pans are equal. Here, the rock has a mass of 600 grams.

A **triple beam balance** has a single pan. Three scaled beams with "riders" are used to measure the mass of what is placed in the pan.

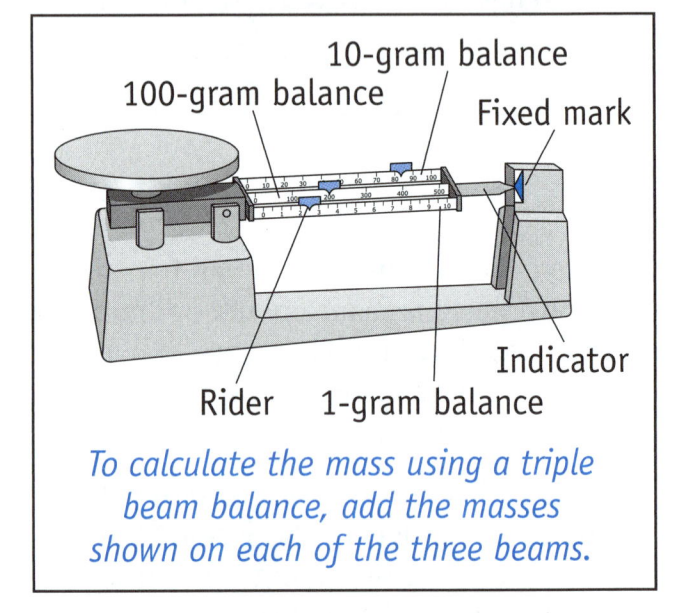

★ First, move the three riders to the left, so that the balance reads zero. Then place the object to be measured in the pan.

★ Move the 100 gram rider to the right until the indicator is just about to drop below the fixed mark.

★ Then do the same with the 10 gram rider.

★ Finally, move the 1-gram rider to the right until the indicator is lined up exactly with the fixed mark. Now add up the values on all three beams — the 100, the 10, and the 1 — to find the mass.

To calculate the mass using a triple beam balance, add the masses shown on each of the three beams.

USING A GRADUATED CYLINDER

To measure the volume of a liquid, scientists pour the liquid into a **graduated cylinder**. The cylinder usually has lines for each milliliter (mL) up to 100 milliliters (mL). The surface of the water curves up the sides of the cylinder. Measure the volume of the liquid from the flat bottom of the curve, known as the **meniscus**. See which line on the side of the cylinder is closest to the bottom of the meniscus.

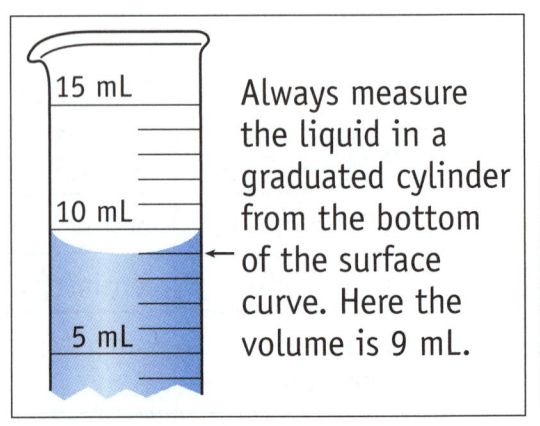

Always measure the liquid in a graduated cylinder from the bottom of the surface curve. Here the volume is 9 mL.

CHAPTER 5: SCIENTIFIC INQUIRY

To measure the volume of a small, irregular solid, like a rock, add water to a graduated cylinder. Record the volume of the water. Now put the solid into the graduated cylinder. Record the new volume. Subtract the original volume of water from the new volume. The volume of a solid object is usually recorded in cubic centimeters. Each milliliter equals one cubic centimeter: $1mL = 1cm^3$. A **cubic centimeter** is the volume of a cube that is one centimeter in length along each side.

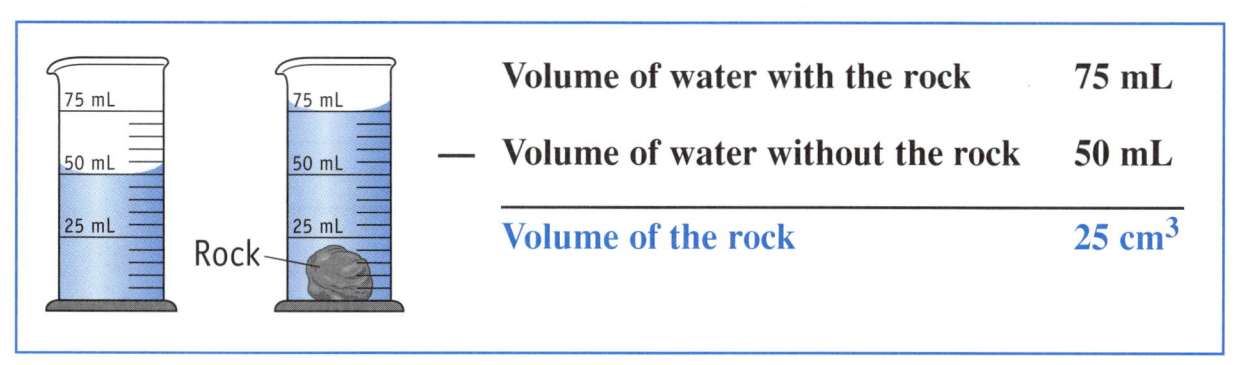

Volume of water with the rock	**75 mL**
− **Volume of water without the rock**	**50 mL**
Volume of the rock	**25 cm³**

APPLYING WHAT YOU HAVE LEARNED

◆ These two graduated cylinders each have the same amount of water. What is the volume of the stone in the second graduated cylinder?

_____ cm³

◆ How did you get that answer?

_____

_____

ANALYZE THE DATA

Once a scientist has gathered data from an experiment, the data must be analyzed. A scientist looks for patterns in the data. The best way to do this is often to organize the data into a table, graph or chart. In this way, a researcher can often see a relationship between two variables.

TABLES

Tables list information in columns and rows. To interpret a table, pay close attention both to the headings for the columns and rows and to the units of measurement that are used. Let's see how this applies to a table by examining the following information:

A group of scientists has been studying the temperature of the atmosphere — the air surrounding Earth — at various altitudes above Earth's surface.

TEMPERATURE OF THE ATMOSPHERE

Altitude (km above sea level)	Temperature (degrees Celsius)
0 km	18°
2.5 km	−2°
5.0 km	−18°
7.5 km	−30°
10 km	−42°
15 km	−52°

In this investigation, scientists measured the temperature of the atmosphere at various altitudes. They recorded their results in the above table.

★ The first column shows the **independent variable** — the altitude (height) of the atmosphere above sea level. This variable was controlled by the scientists. They could measure the temperature at any altitude they chose.

★ The second column shows the **dependent variable** — the temperature of the atmosphere at that particular altitude. The value of this variable *depended* on atmospheric conditions — what the temperature was at that altitude when measured by scientists.

APPLYING WHAT YOU HAVE LEARNED

Use the table above to answer the following questions:

◆ What happened to the temperature of the atmosphere as the altitude increased from sea level to 15 km above sea level? _____

◆ What would you hypothesize or predict the temperature of the atmosphere to be at 17 km above sea level? _____

◆ What would you predict to be the temperature of the atmosphere at 12.5 km above sea level? _____

LINE GRAPHS

A **line graph** shows a series of points on a grid connected by a line. The title tells what the graph shows. Each point on the grid represents a specific quantity. The purpose of a line graph is to show how two or more variables are related.

Usually, a line graph shows how changes in one variable will lead to changes in the second variable.

★ The **X axis** usually represents values for the *independent variable*. The X axis is the bottom line of the graph. As you move to the right on the graph, the value of X increases.

★ The **Y axis** represents values for the *dependent variable*. The left side of the graph shows the Y axis. As you move up this line, the value of Y increases.

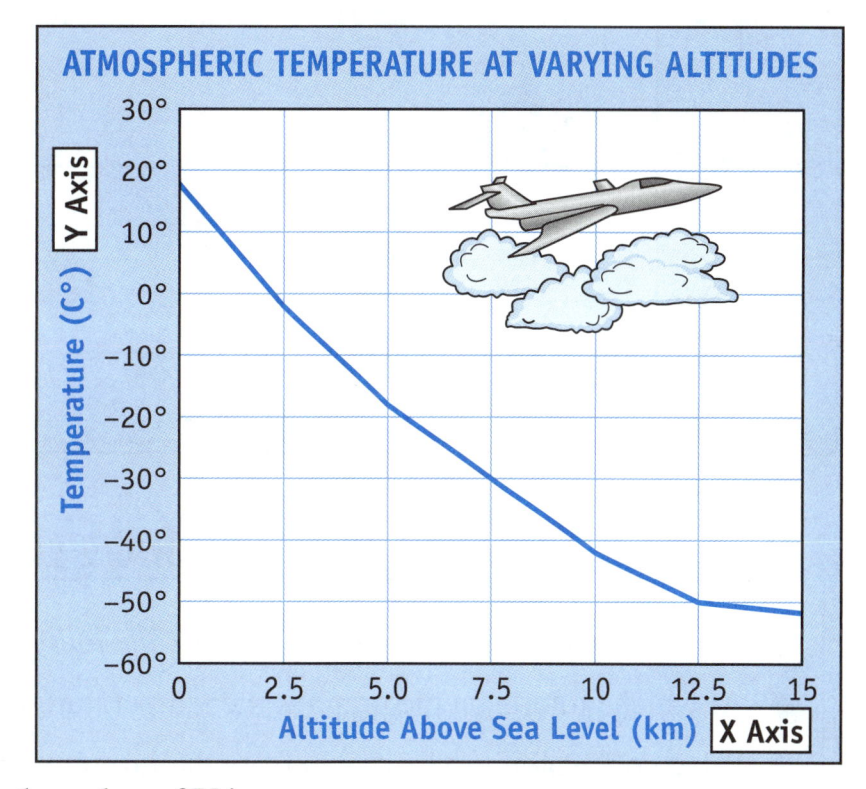

★ The line shows what happens to the value of Y as X changes.

APPLYING WHAT YOU HAVE LEARNED

Use the line graph above to answer the following questions:

The line graph shows information about atmospheric temperatures at different altitudes.

◆ Which is the independent variable? _____

◆ Which is the dependent variable? _____

◆ What relationship do these variables seem to have? _____

BAR GRAPHS

A **bar graph** is made up of parallel bars of different lengths. It is used to make a comparison of different items or of one item at different times. Each bar represents a quantity. Either each bar is labeled, or a key explains what each bar represents. The **Y axis** often indicates what quantities the bars show.

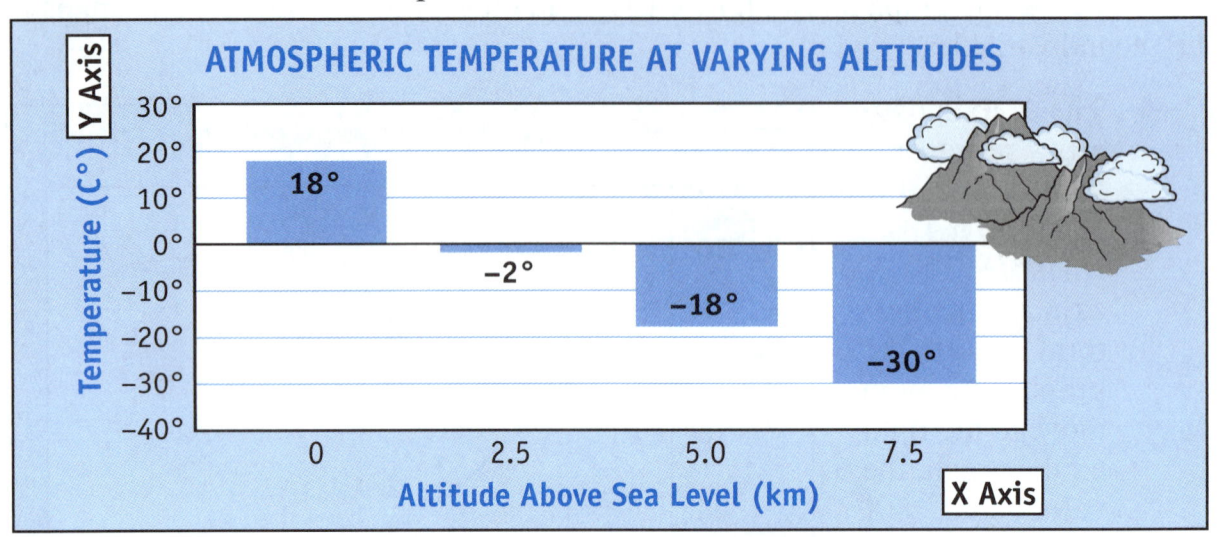

ATMOSPHERIC TEMPERATURE AT VARYING ALTITUDES

Altitude Above Sea Level (km)	Temperature (C°)
0	18°
2.5	−2°
5.0	−18°
7.5	−30°

APPLYING WHAT YOU HAVE LEARNED

Use the bar graph above to answer the following questions:

◆ At which altitude on the graph is the temperature the lowest? _____

◆ What conclusion can be drawn from the information in the graph? _____
_____

◆ How does a bar graph differ from a line graph? _____
_____
_____

◆ Gather a set of data on a scientific topic. Then on a separate sheet of paper create both a line graph and a bar graph using the same data.

FLOW CHARTS

A **flow chart** is a special type of diagram. It shows a series of steps in a process. Each step is placed in some geometric shape or represented by a picture. The shapes or pictures are connected by arrows or lines to indicate the order of steps in the process. Sometimes the process may move forward in several ways. Arrows will indicate when a step may lead to different choices or outcomes.

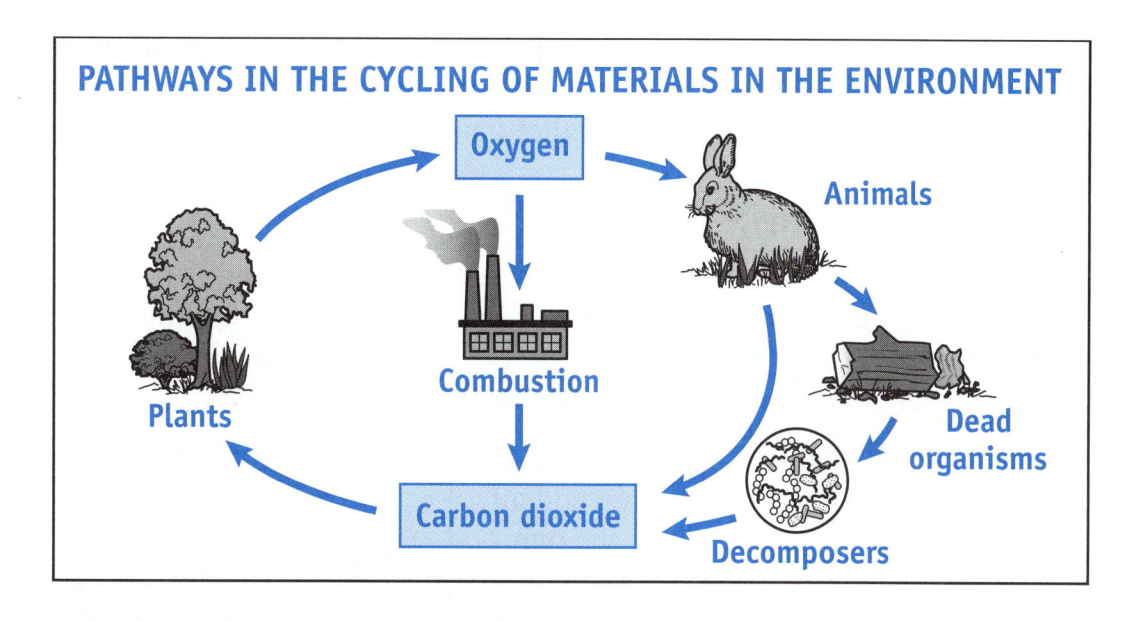

PATHWAYS IN THE CYCLING OF MATERIALS IN THE ENVIRONMENT

APPLYING WHAT YOU HAVE LEARNED

Use the flow chart above to answer the following questions:

◆ What is one source of oxygen in the atmosphere? _____

◆ What happens to oxygen when fuels are burned? _____
_____

INTERPRETING DATA

By organizing their data into tables, graphs, and charts, scientists can see relationships between variables more easily. Two things that scientists often look for is how variables are changing, and what is "typical" for a particular variable.

HOW VARIABLES CHANGE

Scientists often look to see if a variable is increasing, decreasing, or remaining the same. For example, according to the graph on page 53, the higher you travel in the atmosphere — up to the altitude of 15 km — the lower the temperature becomes. This is an example of two variables that are **inversely related**:

> As altitude increases, the temperature decreases.

> As altitude decreases, the temperature increases.

When you look to see how two variables are related, ask yourself if increases in the independent variable (**X**) cause the dependent variable (**Y**) to *increase*, *decrease*, or *remain the same*. This relationship is usually easy to see on a line graph.

Remember, however, that there may be no relationship at all between two variables. As the independent variable (**X**) increases, the dependent variable (**Y**) may remain the same or randomly change without any fixed pattern.

A "TYPICAL" VALUE

Suppose you have the following grades on tests in your science class: **80, 78, 85,** and **97**. What is your average or "typical" grade? Here, your average grade is 85. Sometimes you may do better or worse, but the score you typically receive on a science test is 85. There are several ways scientists determine what is "typical":

★ **Mean.** The **mean** is the **average** value of a variable. To find the mean, add all the values and then divide by the number of values. For example, a scientist has a sample of ten frogs. The frogs weigh 200 g, 200 g, 250 g, 280 g, 300 g 320 g, 320 g, 400 g, 420 g, and 600 g. To find the average, add all ten numbers together (3,290 g), and then divide by 10. The mean is 329 g.

★ **Median.** Often the mean is not representative of a sample since some values are much higher or much lower than the others. In this case, scientists often look at the median. All the values are arranged in order from the lowest to the highest. The **median** is the middle value in this series. In our example, there is an even number of frogs. The median is *halfway* between the two middle values (300 g and 320 g), or 310 g.

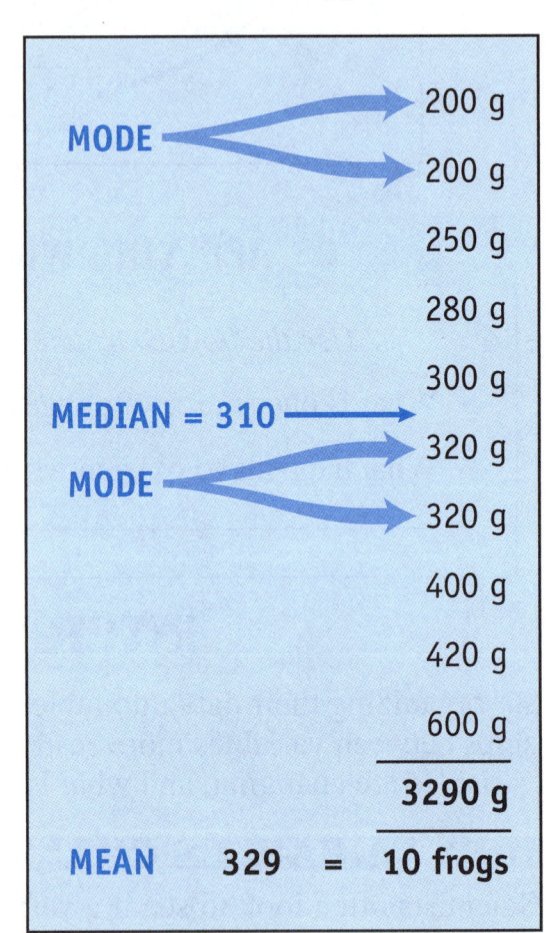

★ **Mode.** A final way scientists measure what is typical is by using the **mode**, or the value that occurs most frequently. The mode may not be unique. Here, two frogs weigh 200 g and two weigh 320 g. In this case, there are two modes: 200 g and 320 g.

```
MODE →  200 g
     →  200 g
        250 g
        280 g
        300 g
MEDIAN = 310 →
        320 g
MODE →
        320 g
        400 g
        420 g
        600 g
        ─────
        3290 g
MEAN  329 = 10 frogs
```

FORM CONCLUSIONS

Scientists draw **conclusions** based on the results of an experiment. These conclusions should relate back to the hypothesis. Often the conclusion states whether or not the hypothesis is valid (*correct*) based on the experimental results.

CHAPTER 5: SCIENTIFIC INQUIRY 57

> ### WHAT MAKES A GOOD CONCLUSION?
> ★ **Logical.** A good conclusion must be logical.
> ★ **Accurate.** A conclusion must correctly report and interpret the data.
> ★ **Supported.** The conclusion should be supported by scientific knowledge and evidence from the investigation.
> ★ **Relevant.** The conclusion should support or reject the hypothesis, or suggest changes in the hypothesis for further study.

The hypothesis is in turn usually related to some scientific theory. Almost every scientific investigation therefore helps to support or reject a theory. Although scientists conduct investigations to test hypotheses related to theories, it is important that they look at the data they collect objectively and not allow their personal **biases** in favor of or against a theory to influence their observation, analysis or conclusions.

More Than One Way to Explain Data. In drawing conclusions, scientists must also recognize that there is often more than one good way to explain a given set of data. A study may show how that a hypothesis could be true, but scientists must be open-minded enough to consider other possibilities. A scientist may hypothesize that as altitude increases, the temperature of the atmosphere drops because the air is thinner. This is true for altitudes up to 17 km above sea level, but in fact changes at even higher altitudes, when the temperature actually increases again.

Scientists Stay within the Evidence. This example shows another danger scientists must take into account when drawing conclusions. Any conclusion that goes beyond the evidence, that misinterprets the evidence, or that follows faulty reasoning should be rejected by the experimenter. In the example under discussion, a researcher cannot assume that temperatures continue to drop at altitudes above 15 km just because they drop from sea level to 15 km above sea level. Such a conclusion would go beyond the evidence and would be incorrect.

APPLYING WHAT YOU HAVE LEARNED

◆ Based on the information shown on the table on page 52, and what you have just read above, what valid conclusions can be reached by the researcher? _____

COMMUNICATE RESULTS

Once a scientist completes an experiment, the results must be communicated to others. Usually this takes place in the form of a written report, article, or oral presentation at a meeting. A good communication presents the methods and procedures followed as well as the results, so that other scientists can *repeat* the experiment and verify the results. This reduces the risk of **bias** and encourages others to build on what was learned. Sometimes disagreements will develop over an experimenter's conclusions. Constructive criticism can help refine an investigator's questions or procedures. Such communication is essential for scientific knowledge to advance.

A good scientist communicates results to other scientists orally or in a written report.

WHAT YOU SHOULD KNOW

☐ You should know that **scientific inquiry** is the process by which scientists ask questions and investigate the natural world.

☐ You should know that there is no single, fixed set of procedures for all scientific investigations. Each investigation determines the procedures needed.

☐ You should know that the following steps are most often used by scientists to conduct an experiment:

    A. The scientist observes the world and **asks a well-defined question**.

    B. The scientist develops a **hypothesis** to answer the question.

    C. The scientist designs an experiment to **test the hypothesis**. Often, an experiment changes one **variable** to see its effect on other variables. Sometimes, however, it is not possible to control all variables.

    D. The scientist chooses appropriate **tools** and **technology** to conduct the experiment, taking **safety** into account.

    E. The scientist **organizes**, **analyzes**, and **interprets the results**. Often there is more than one way the data can be explained.

    F. The scientist **communicates** experimental procedures, results and conclusions to others.

CHAPTER STUDY CARDS

Methods of Scientific Investigation

- **Observe Nature.**
- **Conduct a Laboratory Experiment.** Change one variable to see its effect on another variable.
- **Conduct a Controlled Experiment.** Change one variable for the experimental group but not for the control group.
- **Field Experiment.** Change a variable in its natural setting and observe its effects.
- **Testable Hypothesis.** An educated guess that tries to answer the question under investigation.

Variables

A researcher must have a clear understanding of the variables in an experiment. A **variable** is anything that can **vary**, or change.

- **Independent Variable (X).** A variable that is changed to find how that change affects other variables in the experiment.
- **Dependent Variable (Y).** A variable that is modified as the result of a change to an independent variable in the experiment.
- **Variables Kept Constant.** In a laboratory or controlled experiment, scientists attempt to keep all other variables constant.

Steps in a Scientific Experiment

These steps are often used by scientists, although their order may vary:

- **Ask a well-defined scientific question.**
- **Form a testable hypothesis.**
- **Design an experiment to test a hypothesis.**
- **Select and safely use tools and technology.**
- **Collect data.**
- **Analyze data (table, graph, or chart).**
- **Form conclusions.**
- **Communicate results to others.**

Finding A "Typical" Value

Scientists use several ways to judge what is a "typical" value:

- **Mean.** Average value found by adding all the values and dividing by the number of values.
- **Median.** Found by arranging all the values from lowest to the highest. The middle value (or average of the two middle values if there is an even number) is the median.
- **Mode.** Found by determining the value that appears most frequently in a set of numbers. This value is the mode. There can be more than one mode.

CHECKING YOUR UNDERSTANDING

James hypothesizes that the number of flowers on a plant will increase if the water given to the plant is increased. He wishes to design an experiment to test his hypothesis. James has four pots of geraniums to use in this experiment.

| Plant A | Plant B | Plant C | Plant D |

1. What condition should NOT be the same for the four plants in order to test James's hypothesis?
   A. the size of the plants
   B. the temperature of the water
   C. the number of hours in sunlight
   D. the amount of water they receive

- Examine the Question
- Recall What You Know
- Apply What You Know

SI: A
G7.1

HINT

This question examines your understanding of methods of scientific investigation. You should recall that an experiment usually tests the effects of changing one "independent" variable. In this question, only one variable should be changed: how much water the plants receive. The size of the plants, temperature of the water, and light each receives should be kept the same. Therefore, the correct answer is **Choice D**.

Now try answering some additional questions on your own:

2. Why is an experiment still valuable even when it does not prove that the hypothesis it is testing is true?
   A. It does not need to be repeated.
   B. It shows the experimenter was biased.
   C. It does not need to be communicated to others.
   D. It eliminates the hypothesis as a possible explanation.

SK: A
G6.1

3. What should a student do when heating a solution in a test tube?
   A. cork the test tube
   B. wear safety goggles
   C. point the test tube in any direction
   D. hold the test tube with two fingers

SI: A
G8.1

4. A scientist is conducting an experiment on plant growth. What should the scientist do to find out how tall a plant grew each day?
   A. look at a leaf from the plant under the microscope
   B. put the plant in a sunny place for an hour each day
   C. measure the plant with a ruler at the same time each day
   D. put the plant on a scale each day and weigh it

SI: A
G7.4

Use the following information to answer questions 5 to 8.

A group of students were studying the effects of a new fertilizer on plant growth. They gave the same fertilizer to ten separate pea plants. Each plant was 20 cm high when the experiment began. At the end of six weeks, they measured the heights of the plants again. The table below shows the results of their measurements.

Plant	Height	Plant	Height
Plant A	24 cm	Plant F	28 cm
Plant B	25 cm	Plant G	34 cm
Plant C	26 cm	Plant H	34 cm
Plant D	26 cm	Plant I	36 cm
Plant E	26 cm	Plant J	36 cm

5. What was the mean height of the plants at the end of the experiment?
   A. 28 cm
   B. 29.5 cm
   C. 32 cm
   D. 32.5 cm

6. What was the median height of the pea plants at the end of the experiment?
   A. 26 cm
   B. 27 cm
   C. 28 cm
   D. 29 cm

7. What was the mode of the heights of the plants at the end of the experiment?
   A. 24 cm
   B. 26 cm
   C. 30 cm
   D. 36 cm

8. Based on the experimental results, which conclusion is most valid?
   A. The fertilizer slowed down plant growth.
   B. Fertilized plants grew between 4 cm and 16 cm in length.
   C. Other kinds of fertilizer would make the plants grow faster.
   D. The fertilizer caused the plants to grow faster than without it.

9. Which sentence states a testable hypothesis?
   A. Environmental conditions affect germination.
   B. The available light in the water of a lake is related to its depth.
   C. Boil 100 ML of water, let it cool, and then add 10 seeds to the water.
   D. A lamp, two beakers, and elodea plants are selected for the investigation.

A scientist places five plants of equal size in identical pots. Each plant is placed in a separate room, kept at the same temperature, and given the same amount of water at the same time each day. Each plant is exposed to a different amount of light each day. At the end of 4 weeks, the scientist measures the height of each plant.

10. What is the dependent variable in this experiment?
    A. the growth of each plant
    B. the light each plant receives
    C. the water each plant receives
    D. the number of flowers of each plant

11. A biologist reports a new discovery based on a laboratory experiment. Biologists in other laboratories repeat the same experiment. If the experimental results are valid, what will happen to the other biologists?
    A. They will get similar results.
    B. They will get different results.
    C. They will use different variables to get similar results.
    D. They will use different experimental procedures to get similar results.

12. A group of eighth-grade students conduct an experiment in science class. Each student is shown at work by his or her work station. Which of these students is NOT practicing good laboratory safety?

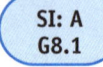

    A.   B.   C.   D.

13. A drug for the treatment of asthma is tested on 1,000 subjects. People in the test are evenly divided into two groups. Members of one group are given the asthma drug, and those in the other group are given a sugar pill. In this experiment, what is the purpose of the subjects who are given the sugar pill?
    A. They act as the control group.
    B. They act as the dependent variable.
    C. They act as the experimental group.
    D. They act as the independent variable.

14. An experimental design includes references to prior experiments, a list of materials and equipment, and step-by-step procedures. What else should be included *before* the experiment can be started?

    A. a set of data
    B. a list of safety precautions
    C. an inference based on results
    D. a conclusion based on the data

    ♦ Examine the Question
    ♦ Recall What You Know
    ♦ Apply What You Know

    SI: A
    G8.1

15. A student is conducting research on the life-cycle of stars. She decides to examine various studies of stars by leading scientists who used different telescopes. Why is it important that she look at the work of others?

    A. She needs to repeat the experiment.
    B. She needs to support her hypothesis.
    C. She needs to learn how to interact with other scientists.
    D. She needs to reduce the risk of bias in her observations.

    SW: B
    G8.2

16. A student wanted to study the feeding habits of young frogs. She placed a group of ten frogs in a glass tank. After feeding the frogs, what should the student do next for safety reasons?

    A. rinse the frogs with cool water
    B. rinse the frog tank with cool water
    C. wash the floor with soap and water
    D. wash her hands with soap and water

    SI: A
    G8.1

17. Almost every experiment begins with a scientist setting out to answer a particular question.

    In your **Answer Document**, create a question a scientist might investigate in an experiment about the health benefits of drinking milk.

    Then, describe how the scientist might change one variable in an experiment designed to answer this question. (2 points)

    SI: A
    G7.3

18. Scientists follow certain safety procedures when carrying out an investigation.

    In your **Answer Document**, identify two safety procedures a scientist should follow in conducting an experiment. (2 points)

    SI: A
    G8.1

19. A student is designing an experiment to test the following hypothesis: salt dissolving in water is a physical change.

    In your **Answer Document**, identify two steps or procedures the student might include in an experiment testing this hypothesis.

    Then, explain how each procedure you identified is important to the final outcome of the experiment. (4 points)

    SI: A
    G6.1

CHAPTER 6

SCIENCE AND TECHNOLOGY

Throughout history, technological advances have greatly influenced the quality of life. In this chapter, you will learn how technology is related to science.

— MAJOR IDEAS —

A. **Technology** is the use of tools, equipment and techniques to make and do things and to extend human capabilities. It is closely connected to science.

B. Technology greatly influences our quality of life. Applications of technology often have both **desirable** and **undesirable consequences**. For example:

★ Automation has made it easier and less costly to make manufactured goods, but this benefit has come at the cost of losing some jobs.

★ Improvements in technology have led to pollution, which threatens the environment and the survival of our planet.

C. **Designing** a product or solution requires consideration of the problem itself and various **constraint**s, such as cost, time for design and production, the available supply of materials, and the potential impact on the environment.

WHAT IS TECHNOLOGY?

Technology is the use of tools and techniques for making and doing things. Technology traces its roots back to the very first humans on Earth. During the Stone Age, people made tools of stone, animal skins and bones to meet their needs. Early humans made arrowheads by chipping away the sides of stones. Then they tied these arrowheads to sticks of wood. They used animal fibers and curved sticks to make bows. They used these simple bows and arrows to hunt animals for food or to fight against one another.

The first humans made arrows by sharpening stones.

CHAPTER 6: SCIENCE AND TECHNOLOGY

With the introduction of farming, human technology became more complex. People learned to plant seeds to grow food. They used new tools to break up the soil, to spread the seeds, and to bring in the harvest. People could now live in one place, so they built permanent homes and towns.

These examples demonstrate how technology extends human abilities to improve the quality of life. The invention of the lever made it possible for people to lift heavy objects. The invention of the wheel helped people move goods more easily. The development of the sailboat made it easier to move goods across lakes and seas. Technology has even helped extend the human senses. The telescope, for example, helped people to see farther than before. The telephone enabled them to listen to others thousands of miles away. Like science, technology has been advanced by the contributions of many different people and cultures. Here are some examples of major turning points:

A.G. Bell's first telephone through which speech was first transmitted electrically in 1875.

MILESTONES OF TECHNOLOGY

Time Period	Technological Advances
Neolithic Revolution (about 8000 BC)	The Neolithic Revolution first occurred in Mesopotamia (now Iraq) and Egypt about 10,000 years ago. People learned to grow crops from seeds and to tame animals and use them for work. These developments allowed people to settle down in one place and establish villages.
Ancient Civilizations (3000 BC–1000 AD)	Early civilizations invented the sailboat and the wheel, mined metals, irrigated fields, built the first cities, and developed calendars and simple forms of writing. Later ancient civilizations, such as Greece and Rome, developed extensive trade routes, and excelled in geometry, architecture, and engineering.
Chinese Achievements (100s–1200s)	The Chinese invented many things that are still in use today. They created paper and invented water clocks to measure the passage of time. They were the first to develop porcelain, silk, the magnetic compass, block printing, and gunpowder.
Hindu Achievements (800s–1000s)	The concept of zero and decimals were invented in Hindu India. Procedures for determining and predicting natural events such as eclipses were known by Hindus centuries before Europe.
Muslim Achievements (800s–1000s)	Muslims developed Arabic numerals, algebra, and the astrolabe for navigation. They preserved learning from Greece and Rome. Their craftsmen excelled in making carpets, glasswork, metalwork, and ceramics. Early Muslim scholars played a key role in advancing astronomy.

Time Period	Technological Advances
Advances in Europe (1500s–1800s)	Many key advances occurred in Europe during this period, including better cannons and guns, telescopes, microscopes, and superior ships. Another invention was the printing press. During the Scientific Revolution and Enlightenment, thinkers developed new scientific methods and calculus.
The Industrial Revolution (Late 1800s–1900s)	The Industrial Revolution began in Great Britain. A series of inventions applied steam power to run machines for spinning and weaving cloth. This allowed British factory-owners to produce more cloth at a much lower price. Steam power was next used to build railroads and steamships. Later this revolution brought steel, the chemical industry, petroleum and electricity to the world.
The Age of Rapid Change (Late 1900s to now)	By the start of the 20th century, there was a major effort to promote scientific research and technology. The result has been a continuous stream of major inventions: the automobile (at the very end of the 1800s), the airplane, radio, television, radar, antibiotics, nuclear energy, missiles, and the computer. Each of these advances has had a major impact on worldwide social and cultural developments.

APPLYING WHAT YOU HAVE LEARNED

As you can see from the above chart, different needs and attitudes have shaped the direction of technological change in different cultures at different times throughout history.

◆ Select one culture from the chart above and explain the importance of its contributions to technology: _____
_____
_____

SCIENCE AND TECHNOLOGY ARE CLOSELY CONNECTED

Technology often involves the application of scientific knowledge to develop tools, materials and processes to help people meet their needs. The growth of scientific knowledge often leads to the development of new technologies that allow people to create better goods and services.

Many of the modern technological wonders we now use in everyday life resulted from scientific research. For example:

★ **Television.** Scientists discovered that the ability of selenium to conduct electricity changed with the amount of light. This discovery led to the invention of the television set.

★ **Microwaves.** A scientist working with an electron tube noticed that a candy bar in his pocket melted. He used this discovery to build the first microwave oven. Because microwaves can penetrate rain, snow, and clouds, these waves are also good for viewing the Earth from space.

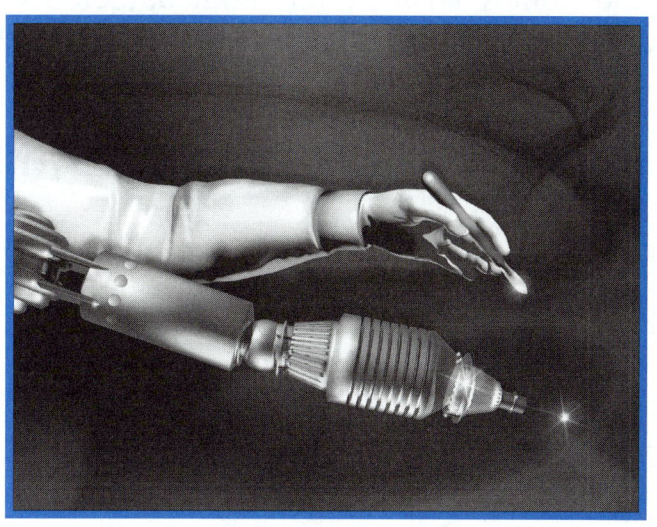

Robotic arms, developed in scientific research, are now able to perform surgery.

At the same time, many scientific advances have occurred because of improvements in technology. The invention of the telescope allowed astronomers to observe distant planets, moons, and stars. The invention of the microscope led to the discovery of microorganisms. The invention of the telephone, computer, and Internet has helped scientists share information more easily.

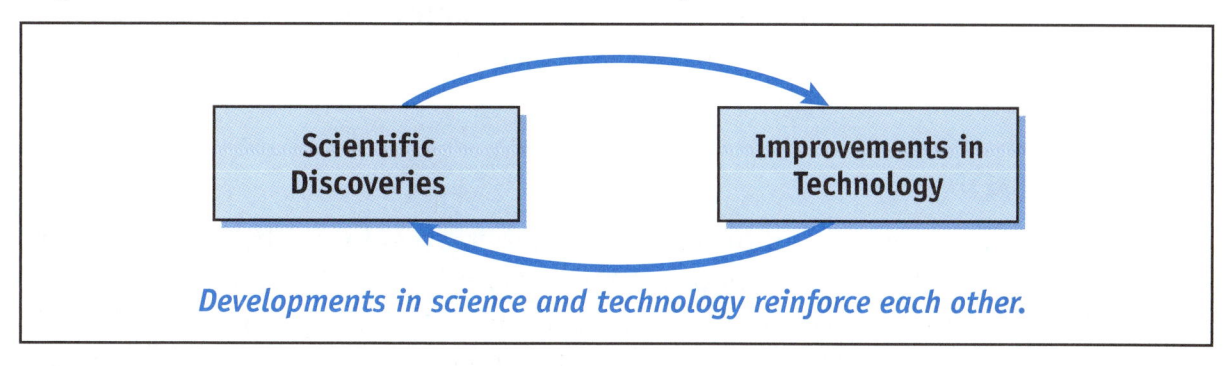

```
   Scientific          Improvements in
   Discoveries         Technology
```

Developments in science and technology reinforce each other.

THE IMPACT OF TECHNOLOGY

The state of technology greatly influences the quality of life in every culture. Today, many Ohioans enjoy a wide variety of foods, warm houses in winter, air conditioning in summer, hot and cold running water, electric lighting at night, cars and airplanes for travel, and CD players, radios, and televisions for entertainment.

But just imagine if you were living in ancient Egypt or with Aztec Indians in Mexico in 1400 A.D. Your life would be dramatically different. Your choice of foods and clothes would be very limited; there would be no hot and cold running water or modern plumbing in your home. There would be no electric lights at night, or cars to take you places. There would be no television or radios to entertain you.

APPLYING WHAT YOU HAVE LEARNED

◆ You have just read about the impact that technology has on everyday life. Describe a typical day in the life of a person your age who lived 1,000 years ago. _____

◆ Now compare one of your own days with the one you just described, highlighting the impact that modern technology has had on your life. _____

THE NEGATIVE EFFECTS OF TECHNOLOGY

Although technology has many desirable consequences — such as strengthening our connections with far-away places, increasing our comfort, and improving our ability to produce goods and services — technology can also have some undesirable effects.

AUTOMATION

Automation is the replacement of human workers by machines. In factories, many jobs once performed by human workers, like the spray-painting of cars, are now conducted by robots. Robots can work all day and night without getting tired or making mistakes. When automation was first introduced, it caused widespread fear among workers. Replacing many workers with automation did in fact lead to the loss of many jobs. For example, millions of telephone operators lost their jobs when they were replaced by automated telephone switchboards.

At the same time, automation has created many new, highly-skilled jobs, such as managing an automated plant, programming computers, repairing robotic equipment, and selling and servicing new products. Today, the automation of manufacturing continues to advance. The result has been a dramatic increase in the general well-being of most people in the developed world.

APPLYING WHAT YOU HAVE LEARNED

◆ Both of these pictures were taken inside a Ford Motor Company plant. They show workers assembling an automobile. The time span is 100 years. How have improvements in technology led to changes in how cars are made?

Ford Motor Plant: 1900

Ford Motor Plant: 2000

_____
_____
_____
_____

WARFARE

Another undesirable effect of technology is that it has greatly increased human destructiveness. In prehistoric times, one group could only attack another with spears, arrows, or clubs. In the ancient world, advances in the introduction of metals led to the use of swords, armor, and chariots. Later, the invention of gunpowder led to the use of cannons and guns.

With each new advance in technology, weapons have become far more destructive. In 1945, the first atomic bombs were exploded over two cities in Japan. Atomic weapons today are far more dangerous and destructive than earlier weapons. These weapons now have the potential to destroy all life on Earth.

The atomic bomb dropped on Hiroshima killed thousands of people.

THREATS TO THE ENVIRONMENT

The development of atomic weapons is not the only technological advance that now threatens life on Earth. Improvements in agriculture, nutrition and health care have led to an explosion of the world's human population. People are using greater amounts of Earth's resources — air, water, and space. People's wastes are polluting Earth's soil and water resources. People are also polluting the atmosphere by burning coal and oil to power their cars and factories, to generate electricity, and to heat their offices and homes. Pollution has led to increasing amounts of carbon dioxide in the atmosphere, bringing about "**global warming**."

APPLYING WHAT YOU HAVE LEARNED

Complete the chart below by giving two examples in each column.

Desirable Effects of Technology	Undesirable Effects of Technology
1.	1.
2.	2.

CONFLICTS AND TRADE-OFFS

The effects of technology sometimes bring economic and environmental goals into conflict. Economic growth may come at the cost of greater pollution. Protecting the environment may slow economic growth. In China, for example, people use portable heaters that burn coal. These heaters make life more comfortable for many Chinese, but they also pollute the atmosphere. In the U.S., many people travel to work by driving alone in cars. Each car burns gasoline, adding pollution to the atmosphere. Economic growth, however, requires these people to drive to work. The environment demands they reduce pollution. These goals appear to be in conflict.

Sometimes new technologies can help people to overcome this dilemma. For example, auto manufacturers are developing cars that use less gasoline. Scientists are also developing new technologies to produce fuels that create less pollution. As the world community becomes more aware of the dangers posed by pollution, people are more willing to devote greater resources to solve the problem.

Geographic location, the availability of resources, and the organization of society often influences what technologies a society finally decides to adopt. For example, in the United States, many people live far from work and are unable to car-pool. They have no choice but to drive to work alone if they want to keep their jobs.

THE TECHNOLOGICAL DESIGN PROCESS

Most new technology results from a process, just as most scientific discoveries result from systematic methods of scientific investigation. In many ways, the process of technological design is similar to the process of scientific inquiry. Technological designs are continually tested, adapted and refined — just like scientific ideas. The steps of the technological design process may vary, just as the steps of a scientific inquiry often do:

STEPS IN TECHNOLOGICAL DESIGN

- **Identify the Problem.** The first step is to identify a human need or problem. The new technology should help to meet this need or to solve this problem.
- **Identify Possible Solutions.** Next, designers need to research the problem to look for possible solutions. They may adopt solutions used elsewhere, or decide why other solutions might not work.
- **Design a Solution.** After carefully studying possible solutions, the designer develops the best solution for this particular problem. The designer often builds a model to test the solution.
- **Evaluate and Test the Solution.** The designer finally reviews the design to see if it would really solve the problem. The designer has to keep in mind the cost of the solution. The designer also judges whether the new technology would create other problems if it were adopted. Sometimes, the design is revised to make sure it solves the problem without creating other new problems. The designer may also show the proposed solutions to other reviewers to get their reactions and ideas. They often consider these factors:
  ★ Does the new technology (*the product or solution*) meet the need?
  ★ How will the technology be manufactured?
  ★ Is the technology easy to operate and maintain?
  ★ How will the technology be disposed of?
  ★ How will the technology be sold?

> ### APPLYING WHAT YOU HAVE LEARNED
>
> ◆ We are surrounded by new technologies, such as computers, cell phones, and i-Pods. Identify one technology and explain what factors led to its adoption. Consider how it is made, sold, and used.
>
> _____
>
> _____
>
> _____

CONSTRAINTS ON TECHNOLOGICAL DESIGN

In creating new technologies, designers often face various **constraints** (*limiting factors*). These constraints affect what scientists and engineers are able to design and produce. Among these constraints are the following:

COST LIMITS

If a designer had an unlimited supply of money, he or she could produce almost anything. However, often a project is worth only a certain amount. The people paying for the project may have only limited funds, or the project may produce only a limited benefit. The costs of design and production should not be more than the benefit the new product or solution provides.

TIME LIMITS

The designer may face a strict deadline: the project must be completed by a certain date. This limits the amount of time the designer has to design a product, build the product, test the product, and examine any problems that might arise.

SUPPLY OF MATERIALS

The supply of materials that the designer has may be limited, or there may be more of one material than another. For example, there may be a large supply of coal but little oil. The designer may try to create a product that requires coal instead of oil. If a critical material is missing, the designer may have to change the product or solution.

ENVIRONMENTAL EFFECTS

As you know, the use of some technologies poses special dangers to the environment. Designers are now very aware of these threats, which act as new constraints. A good designer tries to create products or solutions that solve the problem at hand without harming the environment.

OTHER CONSTRAINTS

Other constraints that designers must take into account include safety, the beauty of the design (*aesthetics*), and the properties of the materials they use. For example, designers must consider how well the parts of an object will fit and interact with other materials. Steel is often used for machine parts because it is both hard and durable. Machine parts are not generally made of glass. A car windshield, however, must be made of glass so that drivers can see while driving. This glass is specially treated with layers of plastic so that it will not splinter if it breaks in an accident.

EVALUATING OVERALL EFFECTIVENESS

After considering the various constraints, designers develop a product or process to solve the problem they want to address. In addition to working around such constraints as a limited budget, a successful design will solve the problem without creating new ones. In evaluating the overall effectiveness of a design, people generally look at how well it solves the problem after it is put into effect.

APPLYING WHAT YOU HAVE LEARNED

Imagine that you are designing a product to help students eat healthier lunches. For example, you may design a new container to keep food fresh. Suppose also that you have **one** constraint: your product cannot be too pricey for students.

◆ What product would you propose? _____

◆ Suppose you were designing the same product, but faced two constraints:
  (1) Your product had to be inexpensive.
  (2) You lived in a community where peanut butter was processed but some people had peanut allergies, and where paper was expensive.

  What product would you propose, based on these new constraints? _____
  _____

◆ How do constraints affect the technological design process? _____
  _____

LIMITATIONS TO SCIENCE AND TECHNOLOGY

People today live in an age of great technological advances. Some people have come to expect that almost every problem can be resolved through the use of science and technology. Despite this hope, you should realize that not every problem can be solved, even with the best science and technology.

APPLYING WHAT YOU HAVE LEARNED

◆ Identify a factor you would consider important in evaluating the effectiveness of a technological design. _____
Explain why this factor is important. _____
_____

WHAT YOU SHOULD KNOW

☐ You should know that **technology** is the use of tools, equipment and techniques to make and do things. Technology is closely connected to science.

☐ You should know that technology influences our quality of life. Applications of technology can have both desirable and undesirable consequences.

☐ You should know that automation has made it easier and less costly to make manufactured goods, but that this has come at the cost of losing some jobs.

☐ You should know that improvements in technology have led to greater pollution, which now threatens the environment.

☐ You should know that designing a product or solution requires consideration of the problem and various constraints, such as cost, time for design and production, the available supply of materials, and impact on the environment.

CHAPTER STUDY CARDS

The Impact of Technology

★ **Technology** is the use of tools and techniques to meet human needs.
★ **Impact of Technology.** Technology can have both desirable and undesirable effects:
  • **Desirable Effects.** Improvements in automation have led to a greater number of goods at lower prices.
  • **Undesirable Effects.** New weapons are far more destructive; the environment is threatened; people have lost their jobs to automation — the use of machinery.

Science and Technology

★ **Science.** Desire to understand the world.
★ **Technology.** Desire to meet human needs.
★ **Science and Technology.** Both reinforce each other.

The Technological Design Process
★ Identify a Problem.
★ Identify Possible Solutions.
★ Design a Solution.
★ Evaluate and Revise the Solution.

Technology and Society

★ Science and technology have advanced through the contributions of many cultures.
  • **Ancient Middle East.** Sailboat, wheel.
  • **Hindu Civilization.** Concept of zero.
  • **Chinese Civilization.** The invention of gunpowder, paper, porcelain.
  • **European Civilization.** Scientific Revolution, Industrial Revolution.
★ Different factors influence which technologies are adapted by a society. These include:
  • Needs, attitudes, and values.
  • Geographic location.
  • Supply of resources.

Constraints on Technological Design

★ **Cost.** There may only be a limited amount of money available.
★ **Time.** There may be only a limited time frame to design and produce the new product or solution.
★ **Supply of Materials.** Only some materials may be available.
★ **Effects on the Environment.** The design should minimize harmful effects to the environment.

CHECKING YOUR UNDERSTANDING

1. Which is an undesirable consequence of increased automation?
   A. New, highly-skilled jobs are created.
   B. Manufacturing can proceed more quickly.
   C. Factories can produce more goods at less cost.
   D. Many jobs requiring manual labor have been lost.

> **HINT:** This question expects you to know that automation is the use of machinery, including computers and robots, to perform tasks. Automation brings many benefits. Automated factories can produce goods much faster and at less cost per item. **Choices A**, **B**, and **C** identify advantages normally associated with automation. Only one choice identifies a disadvantage of automation — that many jobs held by manual workers are lost when they are replaced by machines. **Choice D** is the correct answer.

Now try answering some additional questions on your own:

2. A group of students is designing a product to help people hold water bottles while jogging. Plastic may be more harmful to the environment than paper. Why might the students still prefer to design their product out of plastic?
   A. Plastic would be less expensive than paper.
   B. The students have a limited time to make their design.
   C. Plastic and paper holders could be made equally attractive.
   D. A reinforced paper holder would be just as strong as a plastic one.

Use the information in the box below to answer question 3.

> ★ **Ancient Greece** → **Pythagorean Theorem**
> ★ **Ancient India** → **Decimal Numbers**
> ★ **17th Century France** → **Cartesian Coordinates (X,Y)**

3. What conclusion can be drawn from these examples?
   A. Mathematics has hardly changed since ancient times.
   B. Advances in science and mathematics are often closely related.
   C. Many cultures have contributed to advances in science and mathematics.
   D. Ancient Greeks had little influence on the development of mathematics.

Use the information in the box below to answer question 4.

> Chinese engineers were designing the Three Gorges Dam across the Yangtze River to reduce flooding and to generate electricity. They built a scale model to see how the dam would work before building the actual dam.

4. What constraint made the model most useful to the engineers?
   A. the time it took to build the model
   B. the cost of building the actual dam
   C. the attitudes of the community towards the dam
   D. the possible effects of the dam on the environment

> An engineer is designing a new bridge to cross a river. The local residents are anxious to complete the bridge as soon as possible. They have provided the engineer with plenty of money to complete the project. There are plenty of building materials available in the region.

5. Which constraint should the engineer on this project take most into account?
   A. cost limits
   B. time limits
   C. supply of materials
   D. social considerations

6. A group of designers at XYZ Auto Manufacturing is involved in building the company's new SUV models. The designers have decided to make the vehicle heavier and larger than any of their earlier SUVs. How is the decision of these designers most likely to lead to an undesirable consequence?
   A. There might not be adequate lighting in the factory.
   B. Workers might have to work additional shifts at the factory.
   C. The company may examine design ideas from many sources.
   D. The new vehicle may burn more fuel, which pollutes the environment.

CHAPTER 6: SCIENCE AND TECHNOLOGY   77

Use the information in the box below to answer question 7.

> The nearest star beyond the sun is several light-years away. Scientists would like to send manned spacecraft to this star, but no spacecraft can travel fast enough to reach this destination, even in several thousand years.

7. What does this situation demonstrate?
   A. Stars are moving farther away from each other.
   B. There are limitations to what science can solve.
   C. More money must be spent to educate young people.
   D. Scientists today are not as intelligent as past scientists were.

8. How does the technology a culture possesses affect its scientific knowledge?
   A. Science depends on the way a culture uses energy.
   B. A culture's ideas about nature mainly come from technology.
   C. Scientific knowledge is affected by the tools a culture has available.
   D. People's desire to know about the natural world is rooted in technology.

9. Which has been a desirable effect of technology?
   A. Weapons are now more destructive than ever.
   B. People are more productive at their workplaces.
   C. Many people have lost their jobs to automation.
   D. Greater reliance on fossil fuels is leading to global warming.

10. In developing new products or solutions, designers have to take into account various constraints.

    In your **Answer Document**, describe two types of constraints that designers may face in designing a new product or solution. (2 points)

11. Decisions about developing and using technology can sometimes put environmental and economic concerns into direct conflict with each other.

    In your **Answer Document**, identify one decision to develop or use technology that resulted in a conflict between environmental concerns and economic concerns.

    Then, explain how these economic and environmental concerns are in conflict with each other. (2 points)

12. Science and technology have advanced through the contributions of many different people, cultures and times in history.

    In your **Answer Document**, identify one person, culture or time in history.

    Then explain how some scientific or technological advance was brought about from this person, culture or time in history. (2 points)

CONCEPT MAP FOR NATURE OF SCIENCE

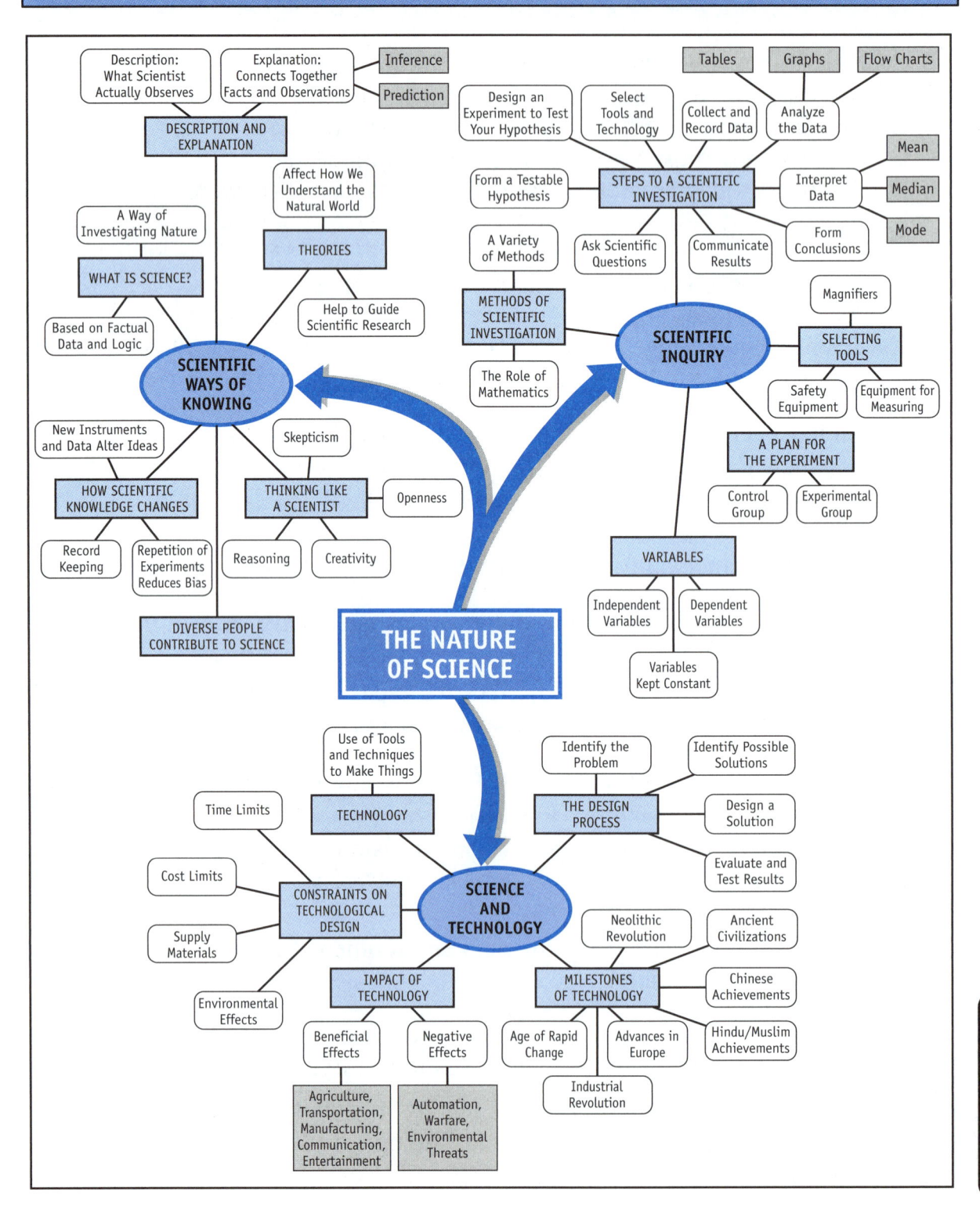

THE NATURE OF SCIENCE

SCIENTIFIC WAYS OF KNOWING

- **WHAT IS SCIENCE?**
  - A Way of Investigating Nature
  - Based on Factual Data and Logic
- **DESCRIPTION AND EXPLANATION**
  - Description: What Scientist Actually Observes
  - Explanation: Connects Together Facts and Observations
    - Inference
    - Prediction
- **THEORIES**
  - Affect How We Understand the Natural World
  - Help to Guide Scientific Research
- **HOW SCIENTIFIC KNOWLEDGE CHANGES**
  - New Instruments and Data Alter Ideas
  - Record Keeping
  - Repetition of Experiments Reduces Bias
- **THINKING LIKE A SCIENTIST**
  - Skepticism
  - Openness
  - Reasoning
  - Creativity
- **DIVERSE PEOPLE CONTRIBUTE TO SCIENCE**

SCIENTIFIC INQUIRY

- **METHODS OF SCIENTIFIC INVESTIGATION**
  - A Variety of Methods
  - The Role of Mathematics
- **STEPS TO A SCIENTIFIC INVESTIGATION**
  - Ask Scientific Questions
  - Form a Testable Hypothesis
  - Design an Experiment to Test Your Hypothesis
  - Select Tools and Technology
  - Collect and Record Data
  - Analyze the Data
    - Tables
    - Graphs
    - Flow Charts
  - Interpret Data
    - Mean
    - Median
    - Mode
  - Form Conclusions
  - Communicate Results
- **SELECTING TOOLS**
  - Magnifiers
  - Safety Equipment
  - Equipment for Measuring
- **A PLAN FOR THE EXPERIMENT**
  - Control Group
  - Experimental Group
- **VARIABLES**
  - Independent Variables
  - Dependent Variables
  - Variables Kept Constant

SCIENCE AND TECHNOLOGY

- **TECHNOLOGY**
  - Use of Tools and Techniques to Make Things
- **CONSTRAINTS ON TECHNOLOGICAL DESIGN**
  - Time Limits
  - Cost Limits
  - Supply Materials
  - Environmental Effects
- **THE DESIGN PROCESS**
  - Identify the Problem
  - Identify Possible Solutions
  - Design a Solution
  - Evaluate and Test Results
- **IMPACT OF TECHNOLOGY**
  - Beneficial Effects
    - Agriculture, Transportation, Manufacturing, Communication, Entertainment
  - Negative Effects
    - Automation, Warfare, Environmental Threats
- **MILESTONES OF TECHNOLOGY**
  - Neolithic Revolution
  - Ancient Civilizations
  - Chinese Achievements
  - Hindu/Muslim Achievements
  - Advances in Europe
  - Industrial Revolution
  - Age of Rapid Change

TESTING YOUR UNDERSTANDING

*At the end of each content unit you will find **Testing Your Understanding**. The purpose of this section is to provide practice questions about the entire unit.*

TEMPERATURE OF LAKE ERIE AT VARYING DEPTHS

Depth in Meters	Temperature in Celsius
1	20°
2	15°
3	10°

1. According to the data on the table above, what is probably the temperature in Lake Erie at a depth of 4 meters?
   A. 15° C
   B. 10° C
   C. 5° C
   D. 0° C

   SI: B
   G8.3

Students placed samples of the same type of bacteria in 10 identical petri dishes with the same nutrients. They placed equal amounts of a chemical in 5 of the dishes and nothing in the other 5 dishes. They checked all 10 dishes under the microscope after five days.

2. What was probably the hypothesis for this experiment?
   A. The chemical would disappear.
   B. The chemical would combine with the nutrients.
   C. The chemical would change the color of the dishes.
   D. The chemical would reduce or eliminate the bacteria.

   SI: A
   G7.3

3. Which is an explanation rather than a description?
   A. The patient broke out in a rash on her neck.
   B. The temperature of the patient was over 102°F.
   C. The patient slept more than twelve hours last night.
   D. The patient's illness was caused by a bacterial infection.

   SK: A
   G8.1

4. Which shows a trade-off favoring economic growth over environmental concerns?
   A. walking to a marathon instead of driving there
   B. using low-cost windmills to create clean energy
   C. installing solar panels on a roof to provide energy for a home
   D. using coal to run a factory even though it creates air pollution

   S&T: A
   G6.2

5. A piece of laboratory equipment is shown to the right. What can this equipment be used to measure?
   A. the mass of a powder
   B. the volume of a liquid
   C. the temperature of a flame
   D. the pressure applied by a gas

   SI: A
   G7.4

6. What effect would increasing the sample size have on the results of a controlled experiment?
   A. The results would have less validity.
   B. The results would show increased bias.
   C. The results would become more reliable.
   D. The results would have to be retested more often.

   SI: A
   G8.2

Use the following illustration to answer question 7.

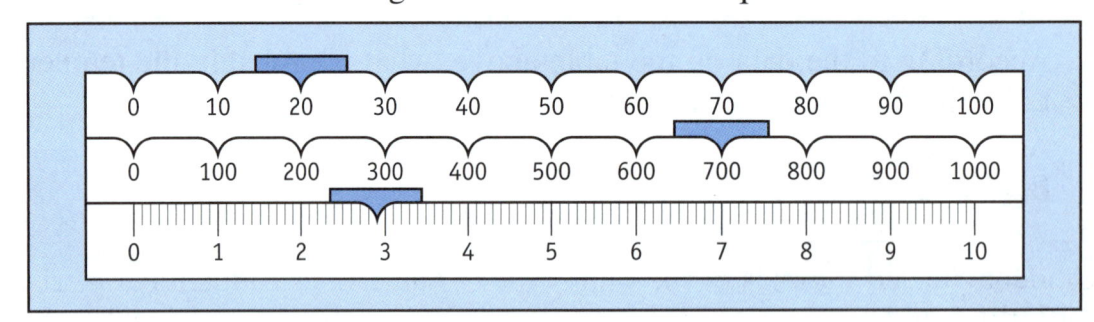

7. The triple-beam balance shown above was used to measure the mass of an object in grams. What mass is being shown?
   A. 723 g
   B. 722 g
   C. 722.9 g
   D. 722.99 g

   SK: A
   G6.2

8. Which is an observation rather than an inference?
   A. The moon has no atmosphere.
   B. Gravity pulls all objects on Earth to the ground.
   C. Some day the universe will probably come to an end.
   D. A solar eclipse occurs because the moon blocks the sun.

   SI: B
   G6.3

9. What has been an undesirable consequence of the adoption of technologies using fossil fuels for energy?
   A. People can travel more easily to distant places.
   B. Many goods can be manufactured at lower prices.
   C. Designs of new technologies often face both cost and time constraints.
   D. Increased carbon dioxide in our atmosphere is causing global warming.

   S&T: A
   G6.2

CHAPTER 6: SCIENCE AND TECHNOLOGY   **81**

Use the following information to answer questions 10 to 14.

A group of students is investigating how long children usually sleep at night. They interviewed ten groups of parents, asking each how long their child slept the previous night. The results of the investigation are shown below:

Subject	Hours Slept	Subject	Hours Slept
Child A	9	Child F	8
Child B	11	Child G	9
Child C	10	Child H	10
Child D	9	Child I	10
Child E	12	Child J	9

10. What is the median number of hours each child slept?
    A. 9 hours
    B. 9.5 hours
    C. 10 hours
    D. 10.5 hours

11. What is the mode number of hours for these results?
    A. 8 hours
    B. 9 hours
    C. 10 hours
    D. 11 hours

12. What is the mean length of time for these results?
    A. 9.0 hours
    B. 9.5 hours
    C. 9.7 hours
    D. 10.1 hours

13. How could the students improve the validity of their investigation?
    A. interview fewer parents
    B. record the favorite foods of each child
    C. specify a precise age for the children in the study
    D. use a different day of the week to conduct the interview

14. In this study, what was the independent variable?
    A. the parents
    B. the children
    C. how long each child slept
    D. when the children went to sleep

---

15. It is very important that scientists keep clear, thorough and accurate records. In your **Answer Document**, provide two reasons why clear, thorough and accurate records are important when conducting an experiment. (2 points)

16. Good scientists try to reduce the amount of bias from entering into their work. In your **Answer Document**, explain why it is important scientists try to prevent bias from affecting their observations.

SW: B
G8.2

Then, explain how the repetition of an experiment is helpful in reducing bias. (2 points)

CHECKLIST OF SCIENCE BENCHMARKS

*At the end of each content unit you will find a **Checklist of Objectives** like the one below. The purpose of these checklists is to help you monitor your understanding of the objectives in the unit before moving on to the next unit.*

Directions. Now that you have completed this unit, place a check (✔) next to those benchmarks you understand. If you are having trouble recalling information about any of these benchmarks, review the lesson indicated in the brackets.

SCIENTIFIC INQUIRY

☐ You should be able to explain that there are different procedures for different types of scientific investigations; procedures are determined by the nature of the investigation, safety considerations and appropriate tools. [Chapter 5]

☐ You should be able to analyze and interpret data from scientific investigations using appropriate mathematical skills in order to draw valid conclusions. [Chapter 5]

SCIENTIFIC WAYS OF KNOWING

☐ You should be able to use skills of scientific inquiry processes (e.g., hypothesis, record keeping, description and explanation). [Chapter 4]

☐ You should be able to explain the importance of reproducibility and the reduction of bias in scientific methods. [Chapters 4 and 7]

☐ You should be able to give examples of how thinking scientifically is helpful in daily life. [Chapter 4]

SCIENCE AND TECHNOLOGY

☐ You should be able to give examples of how technological advances, influenced by scientific knowledge, affect the quality of life. [Chapter 6]

☐ You should be able to design a solution or product taking into account needs and constraints, such as cost, time, trade-offs, properties of materials, safety and aesthetics. [Chapter 6]

UNIT 3: THE PHYSICAL SCIENCES

In this unit, you will learn about the **physical sciences**. The physical sciences study matter, motion and energy.

You can see matter all around you. Matter is anything that takes up space and has mass. Matter is often in motion. To start, stop, or change the motion of matter requires the application of force.

Energy has the ability to move or change matter. Electricity, heat, light and sound are all forms of energy.

Electricity is a common form of energy.

CHAPTER 7: MATTER
In this chapter, you will learn about the structure and properties of matter. You will learn that matter is composed of small particles, and that the arrangement of these particles affects the physical and chemical properties of a substance. You will also learn about physical and chemical changes. In all these changes, the total amount of matter remains constant.

CHAPTER 8: MOTION AND FORCE
In this chapter, you will learn how to describe the motion of an object. You will also learn that all changes in motion are caused by the application of some force.

CHAPTER 9: ENERGY
In this chapter, you will learn about energy. You will learn that energy can take many forms, including potential and kinetic energy. You will discover that one form of energy can transform into another, but that the total amount of energy in such transformations remains constant. Finally, you will learn about both renewable and nonrenewable sources of energy.

CHAPTER 7

MATTER

In this chapter, we will explore the structure and properties of matter.

— MAJOR IDEAS —

A. All **matter** is made up of tiny particles known as **atoms**. Atoms are composed of protons, neutrons, and electrons. The arrangement of these particles affects the properties of a substance.

B. The **physical properties** of matter include its density, shape, state, hardness, and conductivity.

C. Equal volumes of different substances with different **densities** will have different masses.

D. A **physical change** occurs when matter changes its physical properties without changing its structure. A change of state is a physical change.

E. A **chemical change** occurs when substances combine or separate to create new substances with different properties. In a chemical change, the total amount of matter remains the same.

F. The **chemical properties** of a substance refer to its ability to combine with other substances in chemical changes.

Matter is the stuff of the universe. Anything that has **mass** and **takes up space** is known as matter. Matter comes in many different shapes and sizes. For example, air and water are matter. This book is matter. However, not everything is matter. Light and electricity are not matter. This is because they do not have mass and do not take up their own separate space.

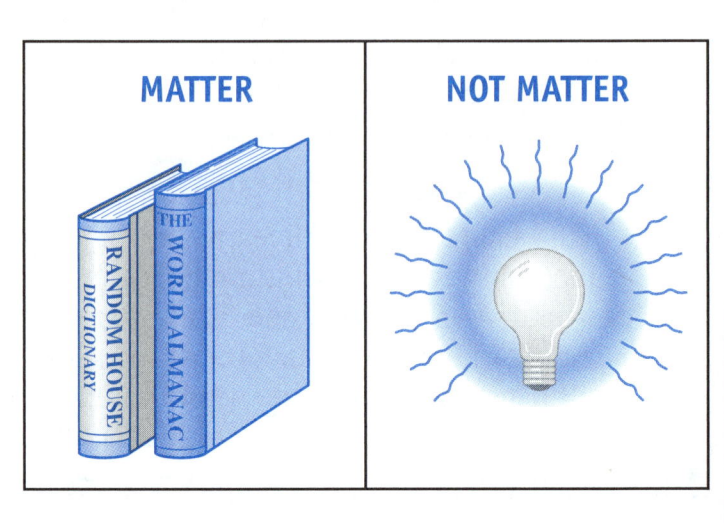

CHAPTER 7: MATTER   85

APPLYING WHAT YOU HAVE LEARNED

★ Identify two examples of matter and two things that are not matter.

Matter	Not Matter
1. _____	1. _____
2. _____	2. _____

THE ATOM

All matter is made up of tiny particles called **atoms**. These atoms are the basic building blocks of matter. These atoms can be broken up into even smaller particles:

★ **Protons.** A **proton** is a particle with a **positive** (+) electrical charge, located in the **nucleus** of the atom. The **nucleus** is the dense center of the atom.

★ **Neutrons.** A **neutron** is a particle with no electrical charge, also located in the nucleus.

★ **Electrons. Electrons** are small particles with a **negative** (-) electrical charge. They move around the nucleus at very high speeds. Compared to protons and neutrons, electrons have no significant mass. Scientists believe electrons move in different **energy levels** around the nucleus. The arrangement of these electrons affects the way each type of atom reacts with other atoms.

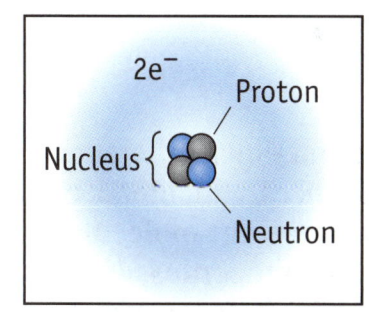

A model of an atom showing electrons moving around the nucleus in electron clouds.

The diagrams below show the structure of two types of atoms. In reality, the electrons are much smaller and farther from the nucleus than shown. If the nucleus were the size of a marble, the electrons would be a football field away.

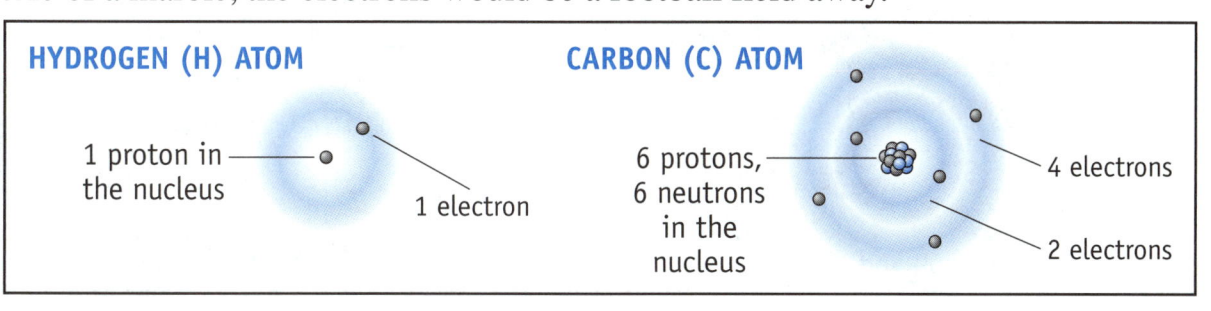

HYDROGEN (H) ATOM

1 proton in the nucleus — 1 electron

CARBON (C) ATOM

6 protons, 6 neutrons in the nucleus — 4 electrons, 2 electrons

ELEMENTS AND COMPOUNDS

There are two types of substances: **elements** and **compounds**. Let's look at each:

ELEMENTS

An **element** is any form of matter made up of one type of atom. Elements have different properties based on their *atomic structure* — how their protons, neutrons, and electrons are arranged. Two important types of elements are metals and nonmetals.

METALS

Metals make up the largest group of elements. Metals are generally solid at room temperature. They are also shiny. Metals are good conductors of electricity and heat. They can be bent or shaped easily, so they can be drawn into wires or hammered into thin sheets.

NONMETALS

Many nonmetals, such as oxygen and nitrogen, are found in nature as gases. Nonmetal solids, like carbon and sulfur, are brittle. They do not bend or twist easily, and do not conduct electricity. Nonmetals are not shiny and do not reflect light.

COMPOUNDS

Sometimes different atoms combine in fixed proportions to make **compounds**: substances with atoms of two or more types, chemically joined together. These atoms often form **molecules** — groups of atoms that share electrons. A compound has its own properties, different from the elements that combined to make it. For example, water (H_2O) is a compound made up of atoms of hydrogen (H) and oxygen (O).

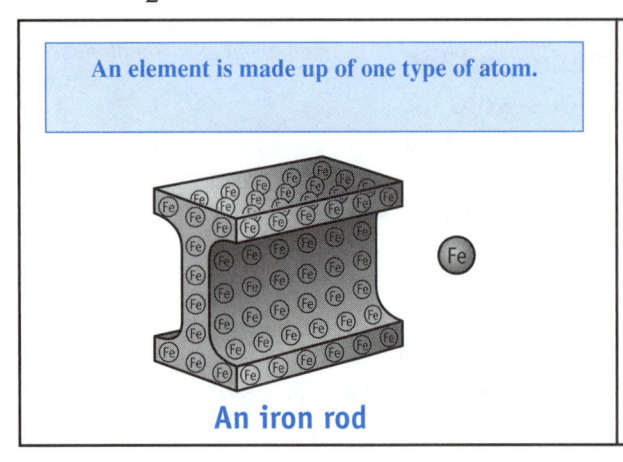

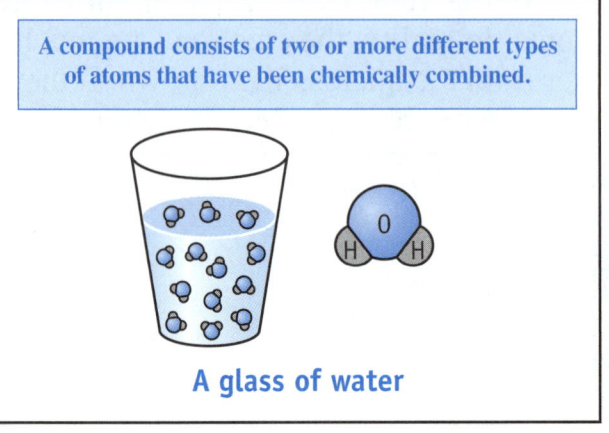

An element is made up of one type of atom.	A compound consists of two or more different types of atoms that have been chemically combined.
An iron rod	A glass of water

THE PROPERTIES OF MATTER

Every substance has its own set of characteristics or **properties** that help scientists set it apart from other substances. The properties are the characteristics scientists use to identify or describe it. For example, when we describe water as "wet" or gold as "shiny," we are describing the properties of these materials. The properties of any piece of matter can be grouped into two types — *physical* and *chemical*.

PHYSICAL PROPERTIES

Pick up an object and you often notice certain things about it — how much it weighs, how hard it is, and what it looks like. These are examples of its **physical properties**. Every piece of matter has its own physical properties. These include its size, shape, color, odor, density, hardness, and how well it conducts heat and electricity.

CHEMICAL PROPERTIES

Chemical properties refer to the ability of a piece of matter to react with other substances. For example, some substances can easily catch fire. If they are heated with oxygen, they will burst into flames. The ability of a substance to combine with oxygen is a chemical property. Hydrogen gas can burn, but water cannot. This means hydrogen and water have different chemical properties.

In this experiment, students mix various chemicals.

APPLYING WHAT YOU HAVE LEARNED

♦ How do the physical properties of a substance differ from its chemical properties? _____
_____
_____

APPLYING WHAT YOU HAVE LEARNED

♦ List two properties of water (H_2O). Then identify each property as either physical or chemical.

Property of Water	Physical or Chemical?
1. _____	1. _____
2. _____	2. _____

Now that you know the basic difference between physical and chemical properties, let's look more closely at some important physical properties.

THE THREE STATES OF MATTER

One important physical property of every piece of matter is its **state** — whether it is a *solid*, *liquid*, or *gas*.

★ **Solids.** Scientists believe that all atoms and molecules are in constant motion. In a solid, these particles are locked into fixed positions. This gives the substance a fixed volume and shape. Its particles still move, but they vibrate in place.

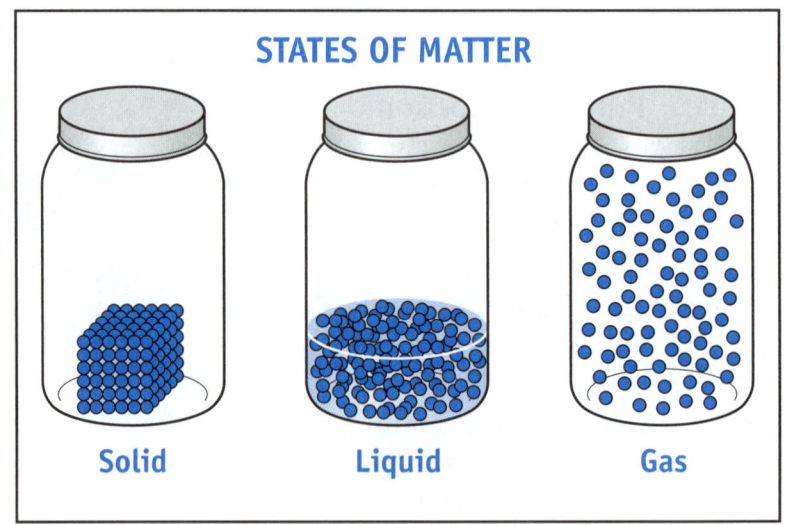

STATES OF MATTER

Solid | Liquid | Gas

★ **Liquids.** When thermal energy (*heat*) is transferred to a solid, its particles begin to vibrate more rapidly. Eventually, its atoms or molecules vibrate so strongly they change their position and start to move around each other. The solid melts and becomes a liquid. Because the particles of a liquid can easily move around each other, the liquid will take the shape of whatever container it is in. A liquid has no fixed shape, but it still has a fixed volume.

Changes of State
1. Solid to liquid
2. Liquid to solid
3. Liquid to gas
4. Gas to liquid
5. Solid to gas

★ **Gas.** If thermal energy (*heat*) is applied to a liquid, its atoms and molecules begin to move around even more rapidly. Eventually, they break all connections with each other and spread out in all directions as a gas. A **gas** has no fixed shape and no fixed volume.

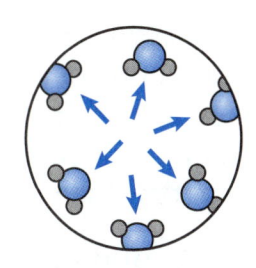

APPLYING WHAT YOU HAVE LEARNED

◆ How do solids, liquids and gases differ from each other? Complete the missing information in chart below.

Characteristic	Solid	Liquid	Gas
Movement of particles	Particles vibrate in place		
Shape			Has no fixed shape
Volume		Has a fixed volume	

DENSITY

Another physical property of matter is its **density** or degree of "compactness." This will depend on both its volume and mass.

★ **Mass** is how much matter something contains. It is usually measured in grams or kilograms.

★ **Volume** is how much space a piece of matter occupies. It may be measured in liters (L), millimeters (mL) or cubic centimeters (cm^3). A cubic centimeter is a cube of 1 cm on all three sides.

These two boxes have the same volume. Assuming each ball has the same mass, the box with the larger number of balls has greater density.

★ The **density** of an object is found by dividing its mass by its volume. The more closely packed together or heavier the individual particles of a substance are, the higher its density it will be. Scientists often measure density in grams per cubic centimeter (**g/cm³**).

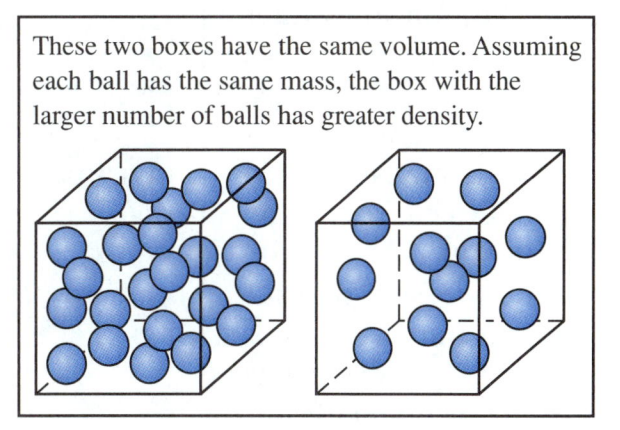

$$\text{Density} = \frac{\text{mass}}{\text{volume}}$$

Because different forms of matter have different densities, the same volumes of matter do not always have the same mass. Equal volumes of different substances usually have **different masses**. A dense substance will have more mass in a given space than a less dense one. For example, lead is more dense than water. One cubic centimeter of lead will have more mass than one cubic centimeter of water.

A solid that is less dense than a liquid will **float** in the liquid. For example, a penny will float in mercury because copper is less dense than mercury. The same penny will sink in water because copper has a greater density than water. Water has a density of 1g/cm^3. Anything with a density greater than this will sink in water, unless it is made in a special shape to displace more water, like a boat.

APPLYING WHAT YOU HAVE LEARNED

◆ An object has a density of 1.29 g/cm^3 and a volume of 2 cm^3.
   What is its mass? _____

◆ Which has a greater mass: 10 cm^3 of a substance with a density of 1.4 g/cm^3 or 10 cm^3 of a different substance with a density of 1 g/cm^3? _____
   What is its mass? _____

A TON OF BRICKS AND A TON OF FEATHERS

Could you tell the difference between a metric ton (1,000 kg) of feathers and a metric ton of bricks? A metric ton of either feathers or bricks has exactly the same mass: one metric ton. However, a metric ton of feathers has a volume of about 400 million cm^3, the size of four tractor trailer trucks. A metric ton of bricks occupies one-half million cm^3, the size of a large-screen TV. The bricks are denser than the feathers because a smaller volume of them has the same mass.

CONDUCTIVITY

A third important physical property of matter is **conductivity** — how well it **conducts** (*carries*) heat, sound or electricity.

★ **Heat Conductivity.** Some materials carry heat better than others. In objects made from these materials, heat moves faster from one end of the object to the other. For example, metals conduct heat well. Wood, plastic and rubber do not conduct heat very well. For this reason, pots are often made of metal while handles are made of wood, plastic or rubber.

★ **Sound Conductivity.** Did you know that sound is caused by the vibration of air? Sound vibrations can also pass through liquids and solids. Some materials conduct sound much better than others. For example, metal and wood conduct sound much better than rubber.

★ **Electrical Conductivity.** Electricity, like heat and sound, can also pass through some materials. Some forms of matter are better conductors of electricity than others. Metals conduct electricity well. Other materials, like rubber or plastic, do not conduct electricity at all. For this reason, wires are usually made of a metal, like copper, surrounded by plastic material that does not conduct electricity. Someone who touches the wire will be protected from the electricity running through the wire, which might otherwise cause a shock.

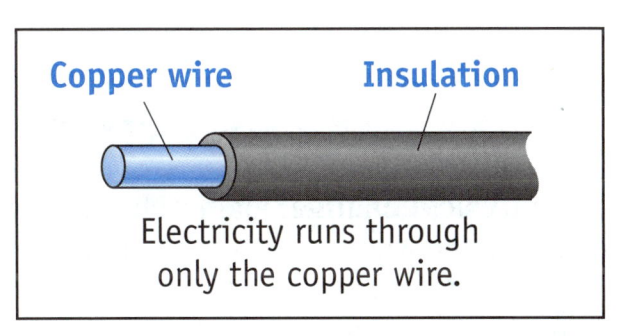

APPLYING WHAT YOU HAVE LEARNED

◆ Look around your house. Identify three objects that are good conductors of either heat, electricity or sound.

(1) _____ (2) _____ (3) _____

◆ Why is it important for engineers and architects to know which materials are good and which are poor conductors? _____
_____
_____

PHYSICAL AND CHEMICAL CHANGES

In the last section, you learned about the properties of matter. In this section, you will learn how matter sometimes changes its properties. For example, a pot of water boils as it changes from a liquid to a gas. This is a **physical change**.

A Chemical Change. Another type of change occurs when a log burns in a fireplace. The log catches fire and burns with crackling flames. After it has finished burning, the log has disappeared. In its place is a heap of ashes. This is an example of a **chemical change**. Physical and chemical changes occur all around us.

PHYSICAL CHANGES

In a **physical change**, matter changes one or more of its physical properties. For example, the water in a pot is a liquid. After the water boils, it becomes a gas. But this substance is still water. It has not changed its basic structure. Each molecule has the same atoms. In fact, *changes of state are always physical changes*. Another physical change occurs when iron is heated. Iron is a very hard substance at room temperature. You cannot scratch it easily. However, if you heat an iron rod to high temperatures, it becomes softer. A blacksmith can shape it into a sword, horseshoe, or tool.

In a physical change, the *chemical properties* of a substance *remain unchanged*. For this reason, physical changes to substances are **reversible** by physical means. If the blacksmith cools down the iron, it will become hard again. Water that has been boiled can also be cooled down and turned back into a liquid.

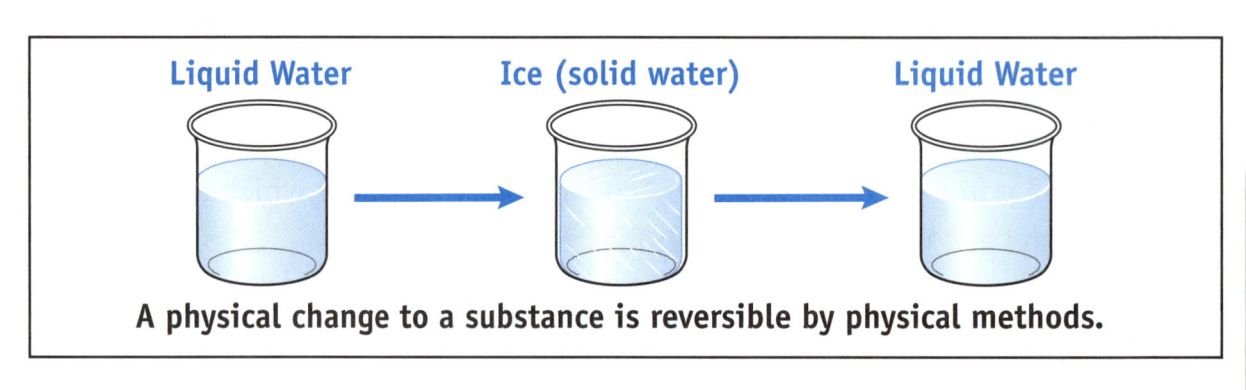

Liquid Water → Ice (solid water) → Liquid Water

A physical change to a substance is reversible by physical methods.

CHAPTER 7: MATTER

CHEMICAL CHANGES

A chemical change is different from a physical change. In a **chemical change**, matter *changes its structure*. One type of matter combines with one or more other types of matter to form a new substance with completely different properties. For example, oxygen gas and hydrogen gas might be combined in a laboratory. When these two gases are brought together, a chemical reaction occurs. Atoms of oxygen and hydrogen join together to form a completely new substance — water. Water (H_2O) has properties that are different from either oxygen or hydrogen gas. At room temperature, water is a liquid, while oxygen and hydrogen are gases.

The materials formed by a chemical change also have new chemical properties. Hydrogen, for example, can burn; water cannot. The rusting of iron is another example of a chemical change. Iron combines with oxygen in the air to create *rust* (iron oxide) — a compound that is different from either iron or oxygen. When you combine chemicals (such as two solids or liquids) and they bubble, this is often a sign of a chemical reaction.

A chemical change results in the creation of new substances.

APPLYING WHAT YOU HAVE LEARNED

◆ Explain how physical and chemical changes differ? _____
_____

◆ Identify one example of each type of change.
A. _____   B. _____

THE CONSERVATION OF MATTER

In a chemical change, the substances that *go into* the reaction and those that *come out* are different. They have different arrangements of atoms with different properties. The total number and type of atoms, however, remain the same *before* and *after* the reaction. They are just combined differently. Because the number of atoms before and after a chemical change stays the same, the total amount of matter remains constant. This principle is known as the **conservation of matter**.

For example, when two hydrogen molecules (**H₂**) are heated with one oxygen molecule (**O₂**), a chemical reaction occurs. They create two molecules of water (2H₂O). This reaction can be shown as follows:

STEP 1: BEFORE THE REACTION	STEP 2: AFTER THE REACTION
Oxygen and hydrogen gases are present in separate molecules (**H₂** and **O₂**).	The oxygen and hydrogen atoms have combined to form water molecules (**H₂O**).

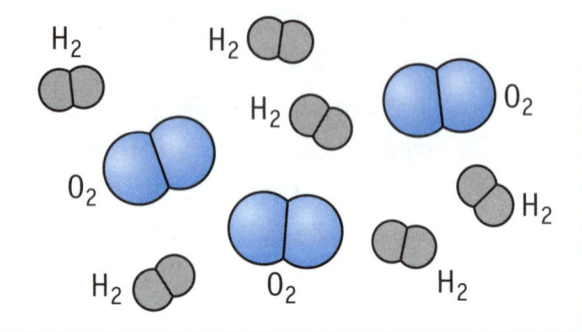

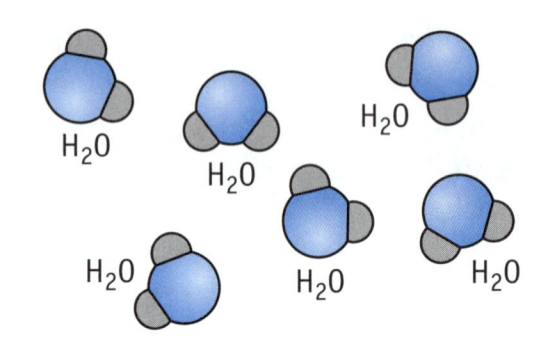

The two gases on the left side are both **elements**: their molecules are made up of the same kinds of atoms. Water is a **compound**: it has different kinds of atoms combined together. Two oxygen atoms and four hydrogen atoms are needed to make every two water molecules. As you can see, the number of atoms stays the same before and after the reaction. The total amount of matter before and after the chemical reaction remains constant.

APPLYING WHAT YOU HAVE LEARNED

A scientist mixed together two common household materials — vinegar and baking soda. The scientist poured one cup of vinegar into a container with one half cup of baking soda. The mixture began to bubble. Soon, bubbles were pouring out of the container. As the vinegar and baking soda combined, they produced carbon dioxide gas.

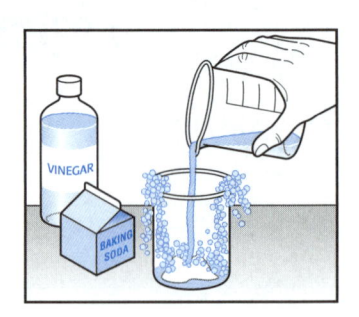

◆ Was this a physical or chemical reaction? _____ Explain your answer.
_____
_____

◆ Many chemical changes occur in daily life. Give two examples of common chemical changes. **(1)** _____ **(2)** _____

MIXTURES AND SOLUTIONS

A **mixture** contains two or more substances that are mixed together *without* being chemically combined. When a mixture is created, *no* chemical change occurs. For example, salt water is a mixture of water and salt. There is no fixed proportion between the ingredients in a mixture. The substances in a mixture often retain many of their original properties — salt water tastes salty. Finally, the substances in a mixture can be separated without a chemical reaction. If salt water is boiled, the water will evaporate and the salt will remain.

A solution is a special type of mixture. In a **solution**, particles of one substance are evenly spread among the particles of another substance. They seem to disappear as they dissolve. This is a physical, not a chemical change. Salt water is a solution. The salt looks as if it has disappeared as it is surrounded by water molecules.

APPLYING WHAT YOU HAVE LEARNED

◆ Explain the differences between a mixture and a compound. _____
_____

WHAT YOU SHOULD KNOW

- [ ] You should know that all **matter** is made up of tiny particles known as **atoms**. Atoms are composed of protons, neutrons, and electrons. The arrangement of these particles affects the properties of a substance.

- [ ] You should know that the **physical properties** of matter include its density, shape, state (*solid, liquid,* or *gas*), hardness, and conductivity.

- [ ] You should know that equal volumes of different substances with different **densities** will have different masses. **Density = mass / volume**.

- [ ] You should know that a **physical change** occurs when matter changes its physical properties without changing its structure. A change of state is a physical change.

- [ ] You should know that a **chemical change** occurs when substances combine or separate to create new substances with different chemical properties. In a chemical change, the total amount of matter remains the same.

- [ ] You should know that the **chemical properties** of a substance refer to its ability to combine with other substances.

CHAPTER STUDY CARDS

Physical & Chemical Properties

Every substance has a unique set of properties that allow scientists to tell it apart from other substances.

★ **Physical Properties.** These include a substance's color, odor, density, state (liquid, solid, or gas), hardness and conductivity.

★ **Density = mass / volume**

★ **Chemical Properties.** The ability of a substance to react with other substances.

Atoms and Subatomic Particles

Atom. All matter is made up of unique particles called atoms. An atom contains:

★ **Protons.** Positively charged; located in the nucleus of the atom.

★ **Neutrons.** Neutral in charge; also located in the nucleus of the atom.

★ **Electrons.** Negatively charged; they move around the nucleus at high speeds; their arrangement greatly affects the properties of a substance.

Physical and Chemical Changes

★ A **physical change** occurs when matter changes one or more of its physical properties, but not its chemical properties. Physical changes are reversible.
 • An example: the melting of ice.

★ A **chemical change** occurs when one type of matter separates or combines with other forms of matter to become something that has a completely different structure.
 • Examples: the burning of wood or the rusting of iron.

Elements and Compounds

★ **Elements.** Any form of matter made up of identical atoms. Examples are carbon (C) and oxygen (O).

★ **Compounds.** A substance made up of atoms of different elements joined chemically together. These elements are always combined in fixed proportions. Examples are water (H_2O) and salt (NaCl).

★ **Molecule.** A group of atoms that share one or more pairs of electrons. For example, water (H_2O) has molecules with oxygen and hydrogen atoms.

CHECKING YOUR UNDERSTANDING

1. Water has a density of 1 g/cm³. Iron has a density of 7.9 g/cm³. Which volume of iron has the same mass as 79 cm³ of water?
   A. 1 cm³
   B. 10 cm³
   C. 100 cm³
   D. 1,000 cm³

   ♦ Examine the Question
   ♦ Recall What You Know
   ♦ Apply What You Know

   PS: A
   G6.1

CHAPTER 7: MATTER   97

> **HINT**
>
> This question tests your understanding of density. Recall that density can be found by dividing the mass of an object by its volume. You should know that the same volume of different substances will usually have different masses because of their different densities. If you know the density of a substance, you can determine its mass by this formula: mass = volume × density. In this example, a volume of 79 cm^3 of water would have a mass of 79g. This is calculated by multiplying 79 cm^3 of water times its density of 1 g/cm^3. What volume of iron would have the same mass? 1 cm^3 of iron would have a mass of 7.9 g; 100 cm^3 of iron would have a mass of 100 × 7.9 g/cm^3 or 790 g; and 1,000 cm^3 of iron would have a mass of 1,000 × 7.9 g/cm^3 or 7,900 g. Thus, the answer is **Choice B**: 10 cm^3 of iron has a mass of 79 g.

Now try answering some additional questions on your own:

2. Why do the same volumes of different substances often have different masses?
    A. They have different densities.
    B. They have different hardnesses.
    C. They have different melting points.
    D. They have different degrees of electrical conductivity.

    PS: A G6.1

3. Which is an example of a chemical change?
    A. A scientist melts ice to create water.
    B. A scientist shapes melted glass into a cylinder.
    C. A scientist grinds an iron rod to make iron filings.
    D. A scientist mixes chlorine gas and sodium to make salt.

    PS: A G6.2

4. Which is an example of a physical change?
    A. A piece of paper burns.
    B. A steel rod is heated until it melts.
    C. Hydrogen and oxygen gases combine to make water.
    D. A plant changes water and carbon dioxide into sugar and oxygen.

    PS: A G6.4

5. Which statement best describes electrons?
    A. They are positively charged particles located in the nucleus.
    B. They are negatively charged particles located in the nucleus.
    C. They are positively charged particles that move around the nucleus.
    D. They are negatively charged particles that move around the nucleus.

    PS: A G7.1

6. What does the density of an object depend upon?
    A. area and height   C. mass and height
    B. mass and volume   D. weight and volume

    PS: A G6.1

7. Which is the best description of an element?
   A. a substance made of one type of atom
   B. a substance made of different kinds of atoms
   C. a substance with positively and negatively charged atoms
   D. a substance with different properties than its individual atoms

   PS: A
   G6.2

Use the following table to answer questions 8–10.

Element	Density	Hardness	Melting Point
Boron (B)	2.34 g/cm^3	9.3	2,076° C
Magnesium (Mg)	1.74 g/cm^3	2.5	605°C
Calcium (Ca)	1.55 g/cm^3	1.75	842°C
Iron (Fe)	7.86 g/cm^3	4.0	1,538°C

8. A scientist collects a sample of one cubic centimeter (1 cm^3) of each of the elements on the table. Which of these volumes will have the greatest mass?
   A. iron
   B. boron
   C. calcium
   D. magnesium

   PS: A
   G6.1

9. Drill bits are often coated to make them harder, so that they can drill through many different types of materials. A coating of which element from the table would make a drill bit the hardest?
   A. iron
   B. boron
   C. calcium
   D. magnesium

   PS: A
   G6.3

10. A group of scientists is studying how elements change from solids to liquids. They create small bricks of boron, magnesium, calcium, and iron. Each brick is 1 cm^3. They place the bricks in an oven and gradually heat them to 1,000°C. At that temperature, which pair of bricks will still be in a solid state?
    A. boron and iron
    B. calcium and iron
    C. magnesium and boron
    D. magnesium and calcium

    PS: A
    G6.3

11. Scientists heat one liter of water until it boils. How do they know this is a physical and not a chemical change?
    A. Bubbles appear in the water.
    B. The water seems to disappear.
    C. The water keeps the same chemical properties.
    D. The temperature of the boiling water remains at 100°C.

    PS: A
    G4.1

CHAPTER 7: MATTER   99

12. Two substances are physically blended together without chemically reacting. They retain their original chemical and physical properties. What is the combination of these substances called?
    A. a mixture
    B. a molecule
    C. an element
    D. a compound

    PS: A
    G6.3

13. How does a compound differ from an element?
    A. Its atoms do not share electrons.
    B. It has a negative electrical charge.
    C. It has atoms of more than one type.
    D. It has the same properties as those of its individual atoms.

    PS: A
    G6.2

14. The density of water is 1 cm^3. Substances whose density is less than that of water will float on the surface of water. Which of the following objects will float on water? (*Hint*: Calculate each density.)

    PS: A
    G6.1

A. a cube with a mass of 45 g, and a volume of 10 cm^3	C. a box with a mass of 55 g, and a volume of 100 cm^3
B. a sphere with a mass of 45 g, and a volume of 15 cm^3	D. a cylinder with a mass of 120 g, and a volume of 60 cm^3

15. What is a substance called that is made up of two or more elements that have been chemically combined?
    A. an atom
    B. an element
    C. a compound
    D. a mixture

    PS: A
    G6.2

16. An element is any form of matter that is made up of only one type of atom.

    In your **Answer Document**, identify two important types of elements. (2 points)

    PS: A
    G6.1

17. All forms of matter undergo change.

    In your **Answer Document**, identify two differences between physical changes and chemical changes.

    Then, provide one example of each type of change. (4 points)

    PS: A
    G6.4

CHAPTER 8

MOTION AND FORCE

In this chapter, you will learn what causes matter to move. You will also learn that all changes in motion are caused by applying some force.

— MAJOR IDEAS —

A. A change in the position of an object is always described in comparison to a reference point.

B. Motion refers to a change in the position of an object over time. Motion consists of both **speed** and **direction**.

C. When an **unbalanced force** is applied to an object, it causes that object to change its motion.

DESCRIBING MOTION

What exactly is motion? **Motion** refers to the change in the position of an object over a period of time. Scientists have developed precise ways of describing and measuring motion:

★ **Change in Position.** A change in the position of an object is always described in comparison to some reference point. The reference point is usually the location of the object at the start of the motion.

★ **Distance.** Distance is the total length traveled by the moving object from the reference point, usually measured in meters (m) or kilometers (km).

★ **Speed.** Speed, or velocity, is the *rate of change*, or average distance traveled by a moving object in a given unit of time, such as meters per second (m/s).

$$\text{Speed} = \frac{\text{Distance Traveled (m)}}{\text{Time Traveled (s)}} \qquad S = \frac{(d)}{(t)}$$

★ **Direction.** Direction describes the location of motion, such as northwards or upwards. It tells *where* the motion is going.

All motion consists of both speed and direction. For example, Mr. Jones drives his car northwards for two hours. During that time, he travels 100 km in the same direction. His motion can be described as follows:

- **Time:** 2 hours
- **Direction:** north
- **Distance:** 100 kilometers
- **Speed:** 50 km per hour

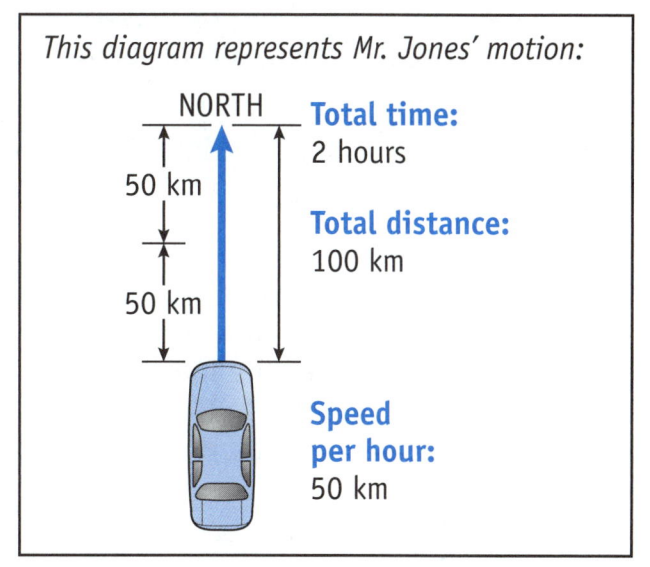

This diagram represents Mr. Jones' motion:

Total time: 2 hours

Total distance: 100 km

Speed per hour: 50 km

In this case, Mr. Jones' motion can also be represented by the following graphs:

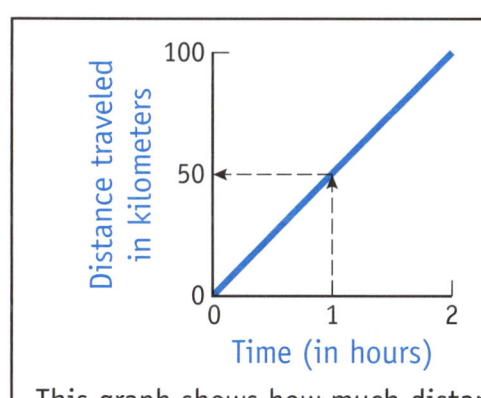

This graph shows how much distance Mr. Jones has traveled. After 1 hour, he has traveled 50 km. After two hours, he has traveled 100 km.

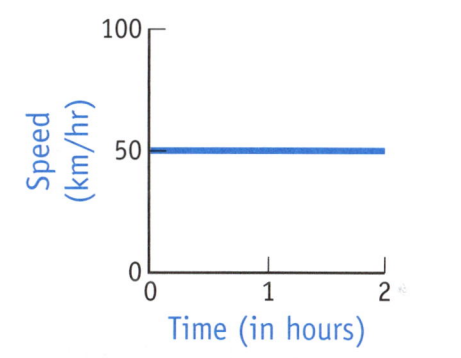

This graph shows Mr. Jones' speed. For the entire trip of two hours, he travels at a constant speed of 50 km per hour.

APPLYING WHAT YOU HAVE LEARNED

◆ Mrs. Jones is traveling on a southbound train to visit her daughter, Mary. Mary lives three hours away by train. The train is traveling 100 km per hour. Demonstrate that you understand motion by filling in the following blanks. Then make a graph on a separate sheet of paper representing her trip. Show time on the X axis and distance on the Y axis.

★ Time: _____   ★ Direction: _____
★ Distance: _____   ★ Speed: _____

TYPES OF FORCE

If a soccer ball is resting on the ground, you can kick it to make it move. This kick applies force to the ball. A **force** is anything that can put matter in motion. In this case, your kick is the force that causes the soccer ball to move. The ball will start rolling, but it eventually slows down. Scientists once thought all objects would naturally return to a state of rest. If they were moving, they would eventually come to a stop.

A foot applies force to a soccer ball.

The Role of Friction. In the 1600s, **Sir Isaac Newton** reached a different conclusion. He believed that objects stop rolling on the ground because of friction. **Friction** is the force caused by the rubbing of two surfaces. Friction resists movement. Newton said that without friction, the soccer ball you kicked would keep rolling forever until some other force stopped it. In fact, that is what would happen if the ball were moving in outer space. Scientists now realize that an object will either remain at rest or remain in motion until some outside force is applied to make it change.

BALANCED FORCES

Sometimes an object is subject to more than one force at the same time. If all the forces acting on an object are perfectly balanced against one another, they will not cause an object at rest to move, or affect the motion of an object that is already moving.

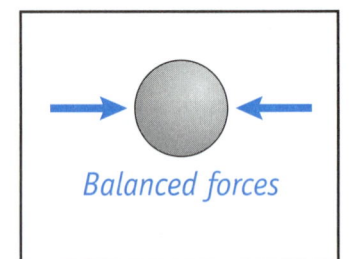

Balanced forces

UNBALANCED FORCES

If the forces are not perfectly balanced, however, more force will act on one side of the object than on the other sides. This **unbalanced force** will change either the speed or the direction of the object, or both. A moving object will speed up, slow down, or change its direction. An object at rest will be put into motion.

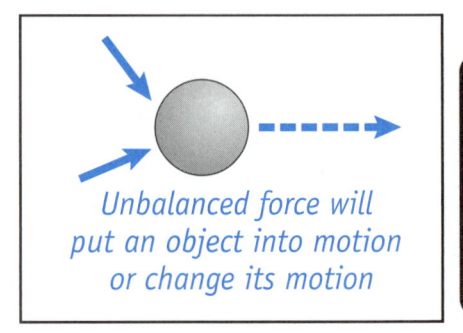

Unbalanced force will put an object into motion or change its motion

APPLYING WHAT YOU HAVE LEARNED

When a car moves ahead, the force from the wheels moving the car forward is greater than the opposing force of friction — the tires rubbing against the roadway.

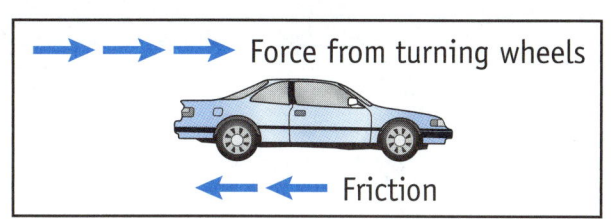

→ → → Force from turning wheels
← ← Friction

◆ Provide two other examples from everyday life in which unbalanced forces lead to changes in an object's speed or direction.

(1) _____

(2) _____

To *increase the speed* of an object in motion, an unbalanced force must give it a "push" in the same direction. To *slow it down*, an unbalanced force must give it a "push" in the *opposite* direction. To get it moving from a state of rest, an unbalanced force must give it a "push" in the direction it is to move. How great must this "push" or force be to change the speed of the object?

★ If the mass of the object is greater, the force must also be *greater*.

★ To make a greater change in speed, the force must be *greater*.

Thus, the force required must be **proportional** to both the mass of the object and the change in speed.

WHAT YOU SHOULD KNOW

☐ You should know that **motion** is a change in position. It is always judged in comparison to a reference point.

☐ You should know that motion can be measured. **Speed** is the average distance traveled by a moving object in a given unit of time, such as m/sec or km/hr. **Direction** describes where the moving object is going.

☐ You should know that the application of an **unbalanced force** to an object will cause it to change its motion. If an object is at rest, it will start to move in the direction of the force. If it is already moving, the force will cause it to change its direction or speed.

CHAPTER STUDY CARDS

MOTION

★ **Distance.** Total length moved by an object (often in meters) in comparison to a reference, or starting, point.

★ **Speed.** Average distance traveled by a moving object in a unit of time, such as meters per second (m/s) or miles per hour.

$$\text{speed} = \frac{\text{distance}}{\text{time}} \qquad S = \frac{d}{t}$$

FORCE

★ **Force.** "Push" or "pull" that causes an object to move or change its motion.

★ **Balanced vs. Unbalanced Forces.**
- A **balanced force** has no effect on an object's motion.
- A force or group of forces that pushes more on one side than the other is an **unbalanced force**. An unbalanced force causes a change in motion.

CHECKING YOUR UNDERSTANDING

1. Which of the following graphs best represents a car moving at a constant speed of 2 m/s?

   PS: B
   G8.2

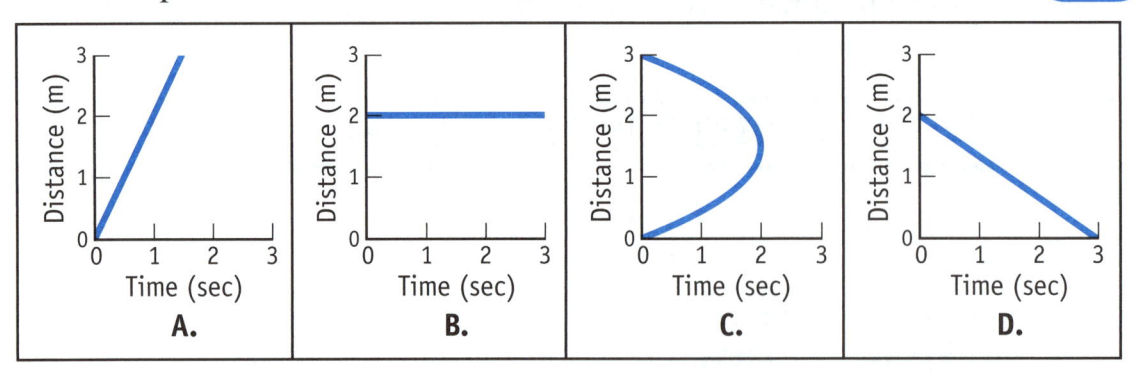

   A. B. C. D.

   HINT: This question asks you to choose the graph that best represents a car moving at a constant speed. The **X axis** shows the number of seconds the car has been moving, while the **Y axis** shows the distance it traveled in that time. **Choice B** shows the same distance after 1, 2, and 3 seconds; **Choice C** shows no travel after 2 seconds; **Choice D** shows no distance at 3 seconds. Only **Choice A** shows the correct distance after 1 and 2 seconds if the car moves 2 meters per second.

Now try answering some additional questions on your own:

2. What will happen to a moving object if all the forces acting on it are perfectly balanced?
   A. The object will increase its speed.
   B. The object will slow down and stop
   C. The object will change the direction of its motion.
   D. The object will continue moving at a constant speed.

   PS: B G8.3

3. A truck travels along a highway for a period of 80 minutes. The graph to the right shows the distance traveled by that truck. According to the graph, during what segment was the truck going at the fastest speed?
   A. Line F
   B. Line G
   C. Line H
   D. Line I

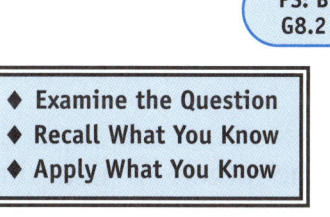

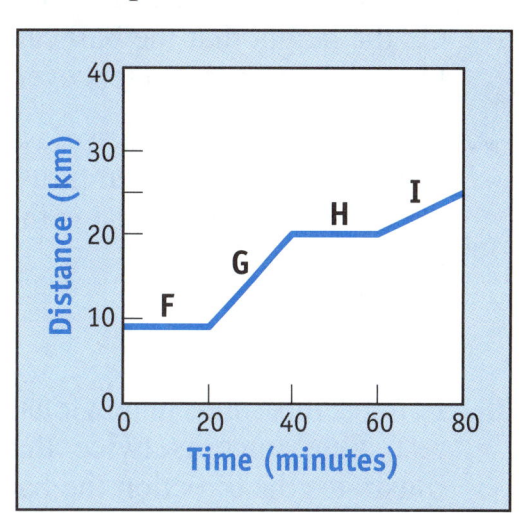

   PS: B G8.2

   ♦ Examine the Question
   ♦ Recall What You Know
   ♦ Apply What You Know

4. A student is looking out of his bedroom window and sees a caterpillar crawling. He decides to time the movement of the caterpillar. The caterpillar crawls from point **A** to point **B** in 5 seconds. What is the caterpillar's speed?

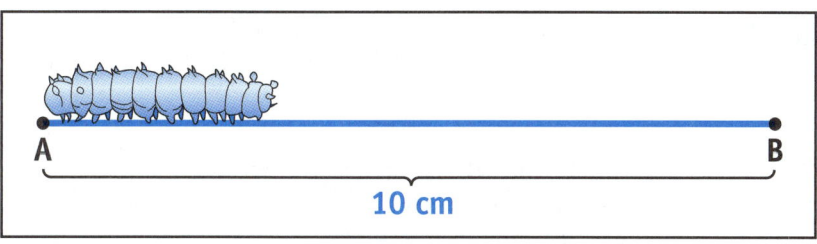

   10 cm

   A. 1 cm/s
   B. 2 cm/s
   C. 5 cm/s
   D. 10 cm/s

   PS: B G8.2

5. Which of the objects on the chart below would require the most force to achieve the same increase in speed on a frictionless surface?

Egg	Basketball	Metal Plate	Glass Vase
10g	100g	200g	400g

   A. egg
   B. basketball
   C. metal plate
   D. glass vase

   PS: B G6.3

6. Students are examining the motion of a basketball as part of an experiment in science class. The diagram to the right shows a constant unbalanced force being applied to the basketball. What is one measurement they could take to describe the motion of the basketball? 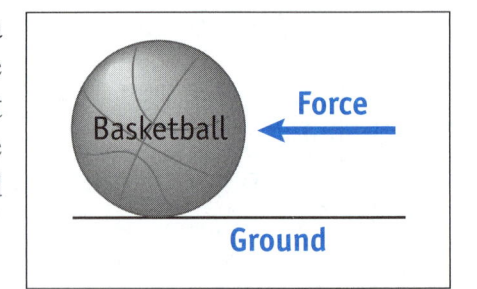
   A. the weight of the basketball
   B. the force applied to the ball
   C. the height that the ball can bounce
   D. the distance it travels every 5 seconds

   PS: B
   G8.2

7. A girl leaves her science class and walks 10 meters north to get a drink at the water fountain. She then turns around and walks another 30 meters south to her mathematics class. What is the distance and direction from her science class to the mathematics class?
   A. 20 m south
   B. 40 m south
   C. 20 m north
   D. 40 m north

   PS: B
   G8.1

8. Force is measured in newtons (N). Two forces are applied to a ball at rest. One force is twice that of the other. Which diagram best illustrates the direction the ball will move?

   PS: B
   G8.3

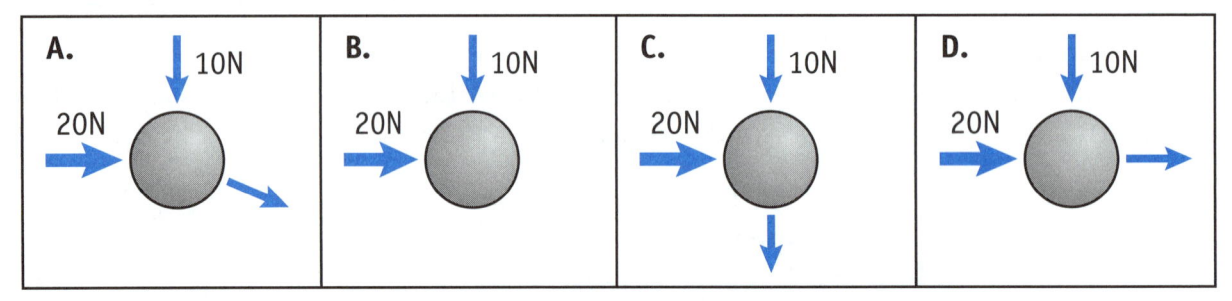

   A. 10N down, 20N right, arrow diagonal down-right
   B. 10N down, 20N right
   C. 10N down, 20N right, arrow down
   D. 10N down, 20N right, arrow right

9. An object is moving at a constant speed. Based on the graph, what is the speed of the object?
   A. 2 m/s
   B. 5 m/s
   C. 10 m/s
   D. 20 m/s

   PS: B
   G8.2

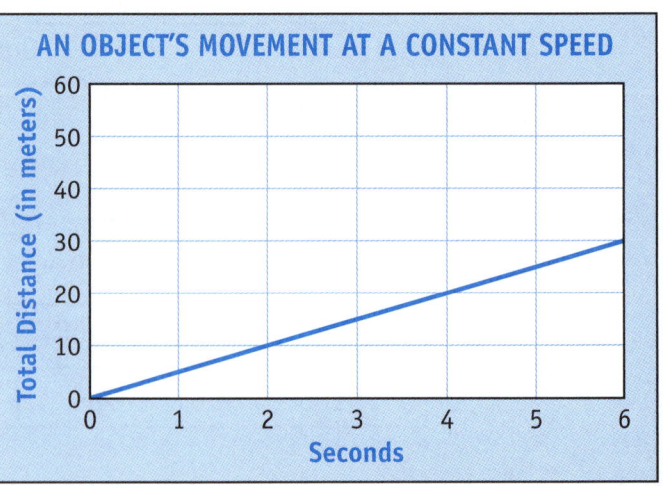

   AN OBJECT'S MOVEMENT AT A CONSTANT SPEED
   (graph: Total Distance in meters vs Seconds; line from (0,0) to (6,30))

10. An unbalanced force can cause an object to change its position or motion.

    In your **Answer Document**, explain how an unbalanced force acts on an object differently than a balanced force does. (2 points)

    PS: B
    G8.3

CHAPTER 9

ENERGY

In this chapter, you will learn about the nature of energy.

— MAJOR IDEAS —

A. **Energy** is the ability to do work.
B. Energy can take different forms. **Kinetic energy** is the energy of movement. An object's kinetic energy depends on its mass and speed. **Thermal energy** is a form of kinetic energy, based on the motion of atoms and molecules. **Potential energy** is the energy stored in an object, based on its position or condition. **Chemical energy** is a form of potential energy. Other forms of energy are *electrical*, *nuclear*, *radiant*, and *acoustic*.
C. Energy can change from one form to another. Even when energy changes from one form to another, it is always **conserved**.
D. **Waves** have energy and can transfer energy through different materials. Vibrations in a material may produce waves that spread away from the source in all directions.
E. Some energy comes from **renewable sources**, such as wind, hydroelectricity, and the sun. Other energy comes from **nonrenewable resources**, such as by burning fossil fuels.

TYPES OF ENERGY

You already know that **force** must be applied to an object to change its motion. **Work** is the application of force over a distance. **Energy** is the ability to do work. There are many types of energy, including:

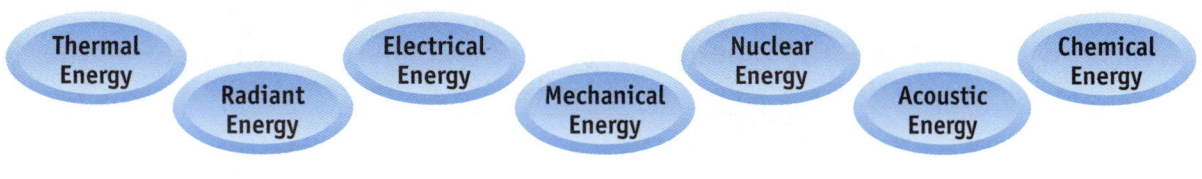

- Thermal Energy
- Radiant Energy
- Electrical Energy
- Mechanical Energy
- Nuclear Energy
- Acoustic Energy
- Chemical Energy

Most forms of energy can be classified as either kinetic energy or potential energy. Let's examine each of these.

KINETIC ENERGY

Any moving object is able to do work: therefore it has energy. This energy of motion is known as **kinetic energy**. When you walk, run or jump, your body exhibits kinetic energy. When water falls, it has kinetic energy. The amount of kinetic energy a moving object has depends on two factors: its *mass* and *speed*. The greater an object's speed, the greater its kinetic energy; the greater its mass, the greater its kinetic energy.

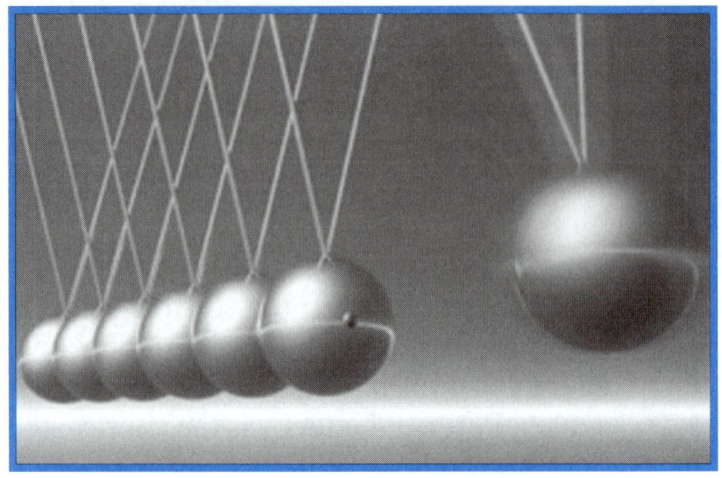

Because this ball is moving, it has kinetic energy and can use this energy to move other objects.

THERMAL ENERGY

Thermal energy is a special form of kinetic energy which we feel as heat. The **kinetic theory** explains that heat is actually caused by the random motion and vibration of atoms and molecules in substances. The thermal energy created by this movement is therefore a form of kinetic energy. **Temperature** is a measure of the speed of movement of these atoms and molecules: it is the average kinetic energy of the particles in an object. An increase in temperature represents an increase in molecular motion. The higher the temperature of an object, the greater the kinetic energy of its atoms and molecules.

APPLYING WHAT YOU HAVE LEARNED

◆ Using the kinetic theory, explain what happens to water as it gets warmer. _____
_____

◆ How do you think the kinetic theory explains changes of state, like the melting of ice? _____
_____

POTENTIAL ENERGY

Potential energy is the energy that an object is able to *store* because of its position or condition. Think of a metal spring. As the spring is pushed down, potential energy is stored in its coils. Once you let go of the spring, it bounces back up, converting its potential energy into kinetic energy.

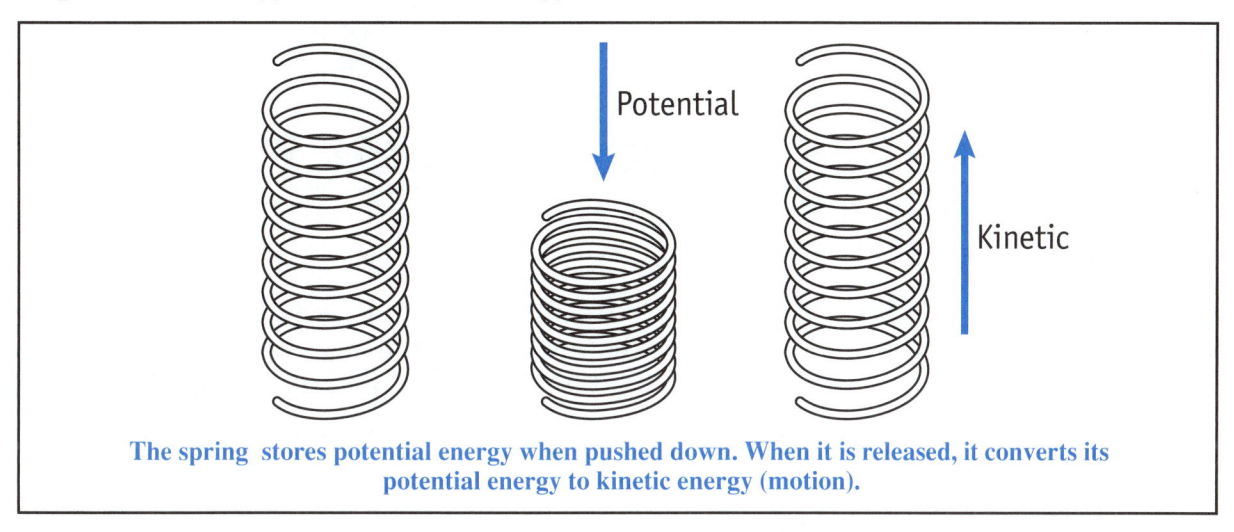

The spring stores potential energy when pushed down. When it is released, it converts its potential energy to kinetic energy (motion).

Stored Potential Energy. One type of potential energy is based on gravity. An object that works against gravity to move into its position actually stores energy. Its position on Earth's surface gives this object an ability to exert force and do work. For example, if a truck is pushed up a hill, it can roll down the hill and pull an object behind it. Energy was "stored" in the truck as it went up the hill. The amount of potential energy in the truck depends on its weight and the distance it was raised.

APPLYING WHAT YOU HAVE LEARNED

◆ Explain how a bucket resting on the edge of a table has potential energy.

_____
_____

CHEMICAL ENERGY

Chemical energy is another form of potential energy. Some types of molecules store energy in the bonds formed by their shared electrons. When these bonds are broken, their energy is released. This happens, for example, when something is on fire. The chemical reaction taking place releases thermal energy and radiant energy. The food we eat, likewise, has stored chemical energy. Our bodies later convert this chemical energy into mechanical and thermal energy.

APPLYING WHAT YOU HAVE LEARNED

◆ How is chemical energy stored and released? _____
_____
_____

OTHER FORMS OF ENERGY

RADIANT ENERGY

Radiant energy is energy, such as *light waves*, *radiowaves* or *solar energy*, that radiates or shines outward from a source. Radiant energy can move through many materials or even through empty space. Radiant energy from the sun travels 93 million miles through space to reach Earth. It is the source of most of the energy on Earth.

ELECTRICAL ENERGY

Another form of energy is **electricity**. Electricity is created by the movement of electrons. These electrons carry negative electrical charges, which can travel through some substances and can even move from one substance to another. Electricity can flow in a circuit and can produce heat, light, sound and magnetism.

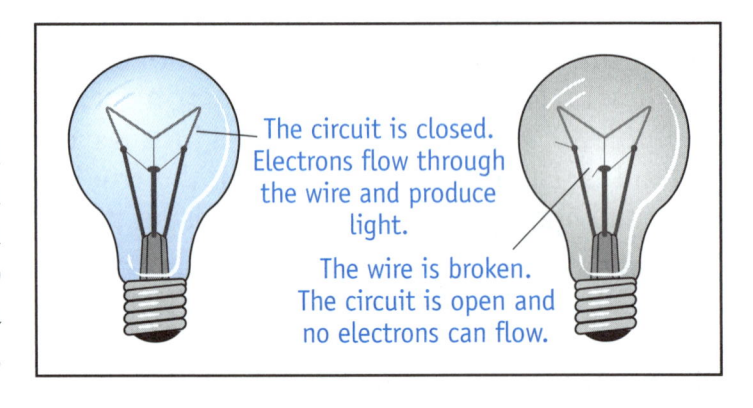

The circuit is closed. Electrons flow through the wire and produce light.

The wire is broken. The circuit is open and no electrons can flow.

NUCLEAR ENERGY

Nuclear energy is another important form of energy. It is the energy stored in the nucleus of an atom. When large nuclei split apart, they release energy. Small amounts of matter are actually converted into immense quantities of energy. Joining the nuclei of smaller atoms together can also release stored nuclear energy. The sun, for example, fuses together the nuclei of hydrogen atoms into helium to produce its energy.

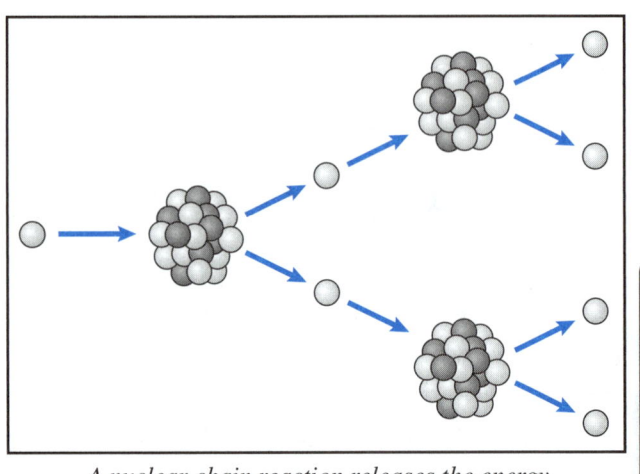

A nuclear chain reaction releases the energy stored in the nuclei of atoms.

CHAPTER 9: ENERGY

APPLYING WHAT YOU HAVE LEARNED

◆ What is radiant energy? _____
_____

◆ What is nuclear energy? _____
_____

THE TRANSFORMATION OF ENERGY

An important quality of energy is its ability to change its form. For example, potential energy can change into kinetic energy. Kinetic energy can change into potential energy. This ability to change form allows energy to move through a substance or to transfer from one substance to another.

A Roller Coaster. Think about a roller coaster ride. The roller coaster car is brought to the top of a hill. It now has stored **potential energy**. As it starts downward, the roller coaster's potential energy decreases. At the same time its **kinetic energy** increases. This kinetic energy then pushes the roller coaster car up the next incline. As it moves upward, its potential energy is again increased, while its kinetic energy decreases as it slows down.

As a roller coaster car loses height, it gains speed, Potential energy is converted to kinetic energy. As it gains height, the car loses speed: the kinetic energy is transformed into potential energy.

A Pendulum. A pendulum provides another example of how potential energy can convert into kinetic energy and then back again into potential energy. To make a pendulum, a weight is attached to a rope, chain, or wire. The weight hangs freely, so that it can swing back and forth.

As the weight is pushed, it starts to swing. As the weight swings downward (from #1 to #2), it picks up speed. Its kinetic energy increases. It reaches its fastest speed when it is at its lowest point, closest to the ground. Then, its speed slows down as it swings up the opposite side (#2 to #3). At the same time, its potential energy increases. The kinetic energy of the pendulum changes into potential energy, until the weight swings up to its highest point (#3). Then it reverses and starts to swing down again, losing potential energy but gaining kinetic energy as it picks up speed.

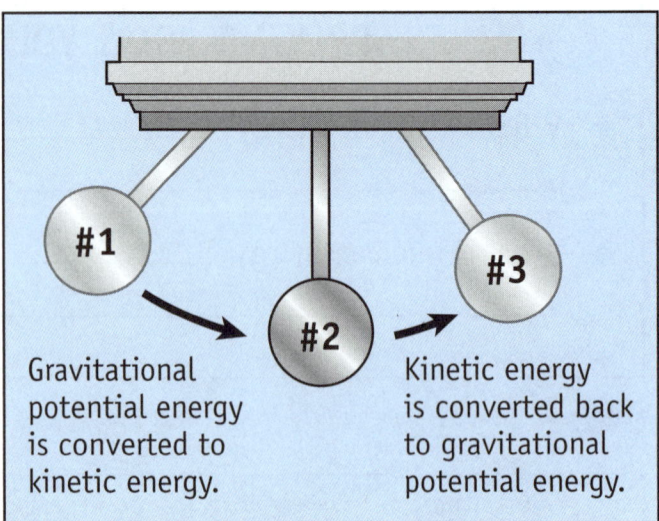

Gravitational potential energy is converted to kinetic energy.

Kinetic energy is converted back to gravitational potential energy.

Each swing of the pendulum is slightly shorter than the last swing.

Other Energy Transformations. Other kinds of energy, besides mechanical energy, can also change their form. For example, chemical energy often converts into thermal energy when a chemical reaction occurs. Other chemical reactions may absorb energy. Photosynthesis is a special kind of chemical reaction in which plants take radiant energy from the sun and convert this into stored chemical energy.

A simple flashlight provides yet another example of energy transformation. Chemical energy from the flashlight batteries changes into electrical energy. The electrical energy runs through a circuit in the flashlight. When it runs through the wires in the bulb of the flashlight, it causes the bulb to glow. The electrical energy has now been transformed into radiant (*light*) energy.

THE CONSERVATION OF ENERGY

Although energy can change from one form to another, it cannot be created or destroyed. This principle is known as the **conservation of energy**. It may seem that energy disappears, but actually it has changed and dispersed. For example, in a fire thermal energy and radiant energy are transferred from the reaction into the surrounding air. Thermal energy heats up molecules in the air, increasing their kinetic motion. The energy of the fire is dispersed but does not go away.

APPLYING WHAT YOU HAVE LEARNED

◆ List two examples from everyday life in which energy changes its form.
   (1) _____ **(2)** _____

◆ On a separate sheet of paper, create a diagram that illustrates the transformation of energy in a flashlight.

◆ Explain what is meant by the "conservation of energy." _____
_____
_____

WAVES OF ENERGY

Some forms of energy spread in special patterns known as waves. A wave is a vibration or disturbance that carries energy through matter or space. The vibration of a material object may cause the wave to start. The waves move away from the source as surface ripples, circles, or spheres.

These waves actually carry energy. For example, if a rock falls into a pond, it brings kinetic energy. This energy passes to the water, which rises up. The water then passes this energy to nearby water, which rises up in turn. The water appears to move but actually stays in place: it is the energy that moves outward in waves.

Seismic (*earthquake*) waves, water waves, sound waves and light waves all transfer energy in this form. Sunlight, for example, travels from the sun to Earth. Its waves carry the energy used by plants for photosynthesis. Infrared waves from the sun carry energy that makes us feel warm in sunlight; ultraviolet waves from the sun bring the energy that can give us sunburn.

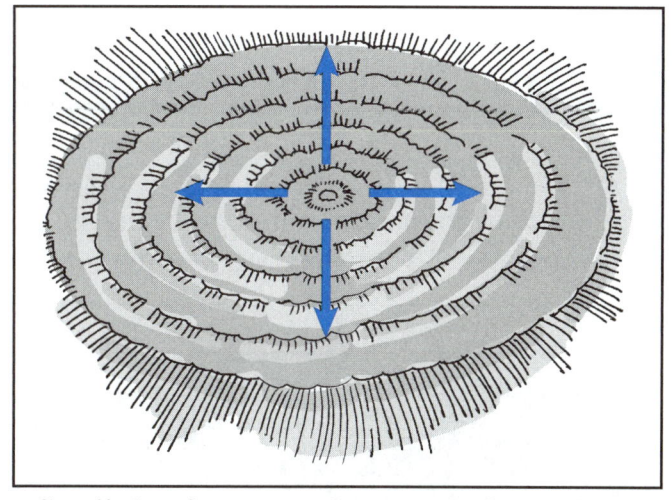

Seismic Waves. Scientists often study how the vibrations in materials cause waves. Seismic waves occur during earthquakes. An earthquake starts when two parts of Earth shift against each other. This sets off a series of vibrations known as seismic waves. The waves spread out from the epicenter, or source of the earthquake. As they travel, they cause the ground to tremble or shake. Underwater earthquakes can cause the ocean to produce giant waves known as **tsunamis**.

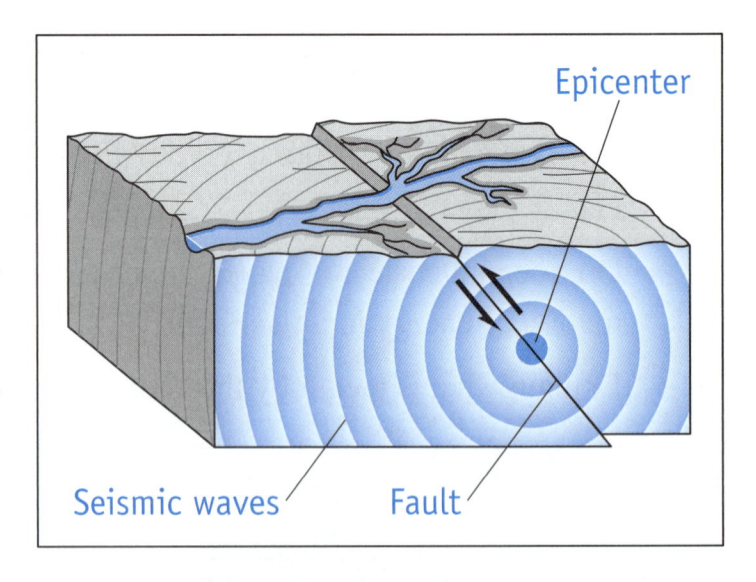

Sound Waves. Vibrations in the air cause what we know as sound. Sound waves always begin with some vibration, such as the movement of our vocal chords. The vibrations then spread out in all directions as waves from the source. They carry **acoustic energy**. Our ears are sensitive to these vibrations, which we hear as sound. Sound can travel through many materials besides air, including wood, metal, and water.

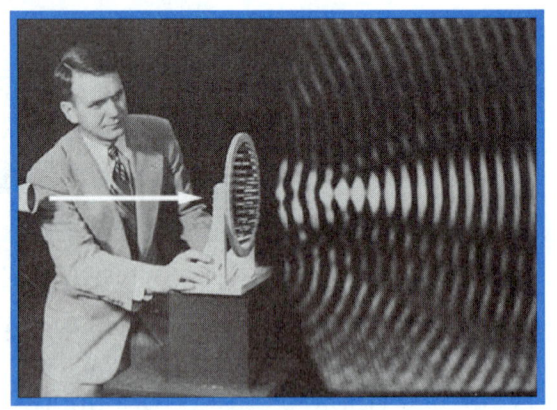

What sound waves look like as sound passes from the horn at left through an acoustical lens.

APPLYING WHAT YOU HAVE LEARNED

◆ On a separate sheet of paper, create a diagram showing how vibrations spread in the air from a glass that falls to the floor and shatters. Be sure to include in your diagram a human ear receiving the sound.

SOURCES OF ENERGY

All around the world, people use energy to maintain their quality of life. We use energy to heat our homes, light our homes, cook our food, power our appliances, and drive our cars. Where does all this energy come from? Scientists have identified two types of energy sources — nonrenewable and renewable.

NONRENEWABLE SOURCES

Many of our energy needs are met from **nonrenewable sources**, such as fossil fuels.

FOSSIL FUELS

Fossil fuels, like coal, oil, and natural gas, are special resources. They have large amounts of stored chemical energy. When they burn, they release this energy. Fossil fuels actually come from the remains of ancient living things.

★ **Coal** is a brown or black rock formed from plants in ancient forests and swamps millions of years ago. After the plants died, they decayed. Over millions of years, heat and pressure changed their remains into coal. Today, we burn coal for electricity and heat. When burned, coal releases the energy stored by plants from the sun many millions of years ago.

★ **Oil and natural gas** are also fossil fuels. They were formed by very tiny plants and animals in the ocean. These living things stored energy, originally taken from the sun through photosynthesis. When they died, they fell to the ocean floor, where mud and sediment covered them. Over millions of years, heat and pressure changed their soft bodies into liquid oil and natural gas.

OIL (PETROLEUM) AND NATURAL GAS FORMATION

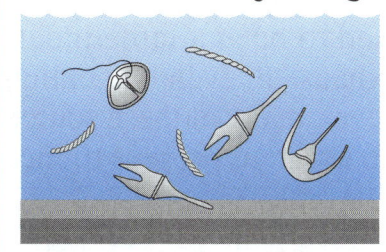

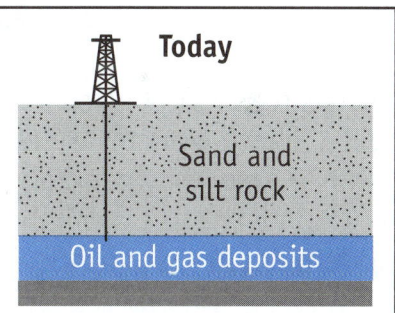

300–400 million years ago	50–100 million years ago	Today
Tiny sea plants and animals died and were buried on the ocean floor. Over time, they were covered by layers of silt and sand.	Over millions of years, the remains were buried deeper and deeper. The enormous heat and pressure turned them into oil and gas.	Today, we drill down through layers of sand, silt, and rock to reach the rock formations that contain oil and gas deposits.

It takes millions of years for fossil fuels like coal and oil to form. They can only be burned once. For this reason, they are considered **nonrenewable**. Although nonrenewable sources cannot be replaced, their use can be extended through careful management. **Conservation** and **greater efficiency** are just two ways to help extend their use.

★ **Conservation.** When we conserve, we use less of something. For example, we conserve energy by not leaving lights on in an empty room, carpooling, lowering the thermostat at night in winter, and using public transportation.

★ **Greater Efficiency.** Scientists are continually improving the efficiency with which energy sources are used. For example, new car engines are able to burn less gasoline than older models to reach the same speeds.

APPLYING WHAT YOU HAVE LEARNED

◆ Identify two ways, besides those above, that people can conserve energy:

(1) _____

(2) _____

◆ Identify another example in which scientists might bring about a greater efficiency in the use of energy sources: _____
_____

RENEWABLE SOURCES

Unlike fossil fuels, some energy sources can be assumed to be available indefinitely. These sources are **renewable** — they cannot be exhausted, even by continuous use. Also, they generally do not create pollution. However, many of these sources are expensive to establish, and many are not powerful enough at present to meet modern energy requirements.

SOLAR ENERGY

Solar energy is energy from the sun. More energy from the sun comes to Earth every day than people can ever use up. Solar energy can be used to heat our homes. Special solar panels, placed in the right location, can absorb sunlight and turn this energy into electricity.

WIND ENERGY

For centuries before the Industrial Revolution, people used windmills to provide the power to grind grain. They also used wind power to drive the sails of sailboats.

A home with solar panels.

Now, windmills are becoming popular again as an inexpensive and clean way to provide energy. Like old-fashioned windmills, today's wind machines use blades to collect the wind's kinetic energy. The wind blows over the blades to make them turn. The blades are connected to a drive shaft that turns an electric generator to produce electricity.

HYDROELECTRICITY

Hydroelectricity is obtained through the movement of water. Waterfalls are used to turn turbines, which create electricity. The kinetic energy of the moving water is then transformed into electricity. Niagara Falls, for example, creates large amounts of electricity. This is a desirable way of producing electricity, since it creates no pollution.

Wind power can produce electricity.

NUCLEAR ENERGY

Every atomic nucleus contains its own energy. A nuclear reactor uses a controlled nuclear reaction to produce heat. The heat produces steam, which drives turbines to produce electricity. **Nuclear energy** might be used to meet our energy needs instead of burning fossil fuels, but there are problems with what to do with nuclear wastes and how to avoid nuclear accidents.

A nuclear power plant.

BIOMASS

Biomass is the leading renewable energy source in the United States today. It is derived from plant-based sources, including paper and wood wastes, ethanol (*a corn-based fuel*), and biodiesel. Although biomass sources produce carbon dioxide when burned, the plants that produced these fuels previously produced oxygen and absorbed carbon dioxide. Thus, the overall impact on global warming is reduced.

GEOTHERMAL ENERGY

Earth's interior is very hot. In some places, cracks in Earth's surface produce steam. Water can also be injected into hot rock. The steam is used to turn turbines to produce electricity. **Geothermal energy** can be used as an efficient heat source, but in some situations the consumer actually needs to be close to the source of heat. Reykjavik, the capital of Iceland, is heated mostly by geothermal energy.

Scientists continue to work on finding new renewable sources of energy. For example, some scientists are using lasers to join together protons to create energy from fusion, like the sun. This would be a cleaner form of energy.

APPLYING WHAT YOU HAVE LEARNED

◆ Describe how electricity is produced from burning fossil fuels, solar energy, water power (hydroelectricity), wind energy, and nuclear energy.
_____
_____
_____

◆ Which renewable energy source do you think shows the most promise for replacing fossil fuels? _____ Explain your answer.
_____
_____

WHAT YOU SHOULD KNOW

☐ You should know that **energy** is the ability to do work.

☐ You should know that energy can take different forms. **Kinetic energy** is the energy of movement. An object's kinetic energy depends on its mass and speed. **Thermal energy** is a form of kinetic energy, based on the motion of atoms and molecules. **Potential energy** is the energy stored in an object, based on its position or condition. Other forms of energy are **electrical**, **chemical**, **nuclear**, **radiant**, and **acoustic**.

☐ You should know that energy can change from one form to another. Even when energy changes from one form to another, it is always **conserved**.

☐ You should know that **waves** have energy and can transfer energy through different materials. Vibrations in a material may produce waves, which spread away from the source in all directions.

☐ You should know that some energy comes from **renewable** sources, such as wind, hydroelectricity, atomic nuclei, biomass, Earth's interior heat, and the sun.

☐ You should know that other energy comes from **nonrenewable** sources, such as by burning fossil fuels.

CHAPTER STUDY CARDS

Forms of Energy
- ★ **Energy.** The ability to do work.
- ★ **Kinetic Energy.** Energy of motion based on the mass and speed of the moving object.
- ★ **Thermal Energy.** A form of kinetic energy based on the vibrations and movements of atoms and molecules.
- ★ **Potential Energy.** Stored energy.
  - One form of potential energy is based on an object's position on Earth and the force exerted by gravity.
  - **Chemical** energy is another form of potential energy.

Transformation of Energy
Energy can change from one form to another. For example:
- ★ **Kinetic energy** can turn into potential energy and back again; examples include a pendulum or roller coaster.
- ★ **Thermal energy** or **chemical energy** can be used to create electrical energy. Electrical energy can be changed into thermal or radiant energy; example: a flashlight.
- ★ **Law of Conservation of Energy.** Energy can change its form, but its total quantity is always conserved.

Other Forms of Energy
- ★ **Electricity.** A form of energy created by the movement of electrons.
- ★ **Nuclear Energy.** When large nuclei split apart or smaller nuclei are joined together they release energy.
- ★ **Radiant Energy.** Energy created by sunlight, radiowaves, or light that radiates from a source.
- ★ **Waves.** Waves carry energy. They can be caused by vibrating certain materials. The waves spread out from the source. Examples: seismic waves, sound waves.

Sources of Energy
- ★ **Nonrenewable.** Nonrenewable sources of energy include fossil fuels (oil, gas, and coal).
- ★ **Renewable.** Renewable sources include:
  - **Solar Energy.** From sunlight.
  - **Wind Energy.** Movement of the wind.
  - **Hydroelectric Energy.** From waterfalls.
  - **Nuclear Energy.** Splitting atomic nuclei.
  - **Biomass Energy.** Comes from plant-based sources.
  - **Geothermal Energy.** From Earth's super-hot interior.

CHECKING YOUR UNDERSTANDING

1. A cement block is dropped from a helicopter. The block falls 500 meters. Which statement accurately describes the falling cement block?
   A. Its potential energy decreases as its kinetic energy increases.
   B. Its potential energy increases as its kinetic energy decreases.
   C. Its potential energy is unchanged as its kinetic energy decreases.
   D. Its potential energy is unchanged as its kinetic energy increases.

PS: D
G7.2

> **HINT**
> This question tests your understanding of potential and kinetic energy. Recall what you know about both forms of energy: then apply this knowledge to the question. As the block falls, its speed increases but its height decreases. Its kinetic energy increases while its potential energy decreases. Thus, the correct answer can only be **Choice A**.

Now try answering some additional questions on your own:

2. What happens when a substance is heated?
   A. The bonds within its atoms break down.
   B. The molecules of the substance slow down.
   C. The molecules of the substance move faster.
   D. The lighter particles in the substance clump together.

   PS: A
   G6.3

3. Which device changes electrical energy into light energy?
   A. a flashlight
   B. a gas stove
   C. an electric oven
   D. a hydroelectric power plant

   ♦ Examine the Question
   ♦ Recall What You Know
   ♦ Apply What You Know

   PS: D
   G7.5

4. A student works after school by mowing lawns. She uses a gasoline-powered lawn mower. As she mows the lawn, the mower gets hot. Which best describes the transformation of energy that occurs?
   A. thermal energy → kinetic energy → nuclear energy
   B. chemical energy → kinetic energy → thermal energy
   C. potential energy → kinetic energy → chemical energy
   D. electrical energy → chemical energy → kinetic energy

   PS: D
   G7.5

5. Which is a form of kinetic energy?
   A. energy of moving molecules in a heated liquid
   B. energy stored in the chemical bonds of a substance
   C. energy holding together the particles in an atomic nucleus
   D. energy stored in an object's position above Earth's surface

   PS: D
   G7.2

6. What do geothermal and nuclear power plants have in common?
   A. They both burn organic materials.
   B. They both produce carbon dioxide.
   C. They both produce hazardous wastes.
   D. They both use thermal energy to turn turbines to make electricity.

   PS: D
   G7.3

7. A group of scientists wishes to examine ways of producing energy that cannot be exhausted. Which energy source should these scientists investigate?
   A. oil
   B. coal
   C. natural gas
   D. geothermal

   PS: C
   G6.8

8. What is the *original* source of the energy found in fossil fuels like coal and oil?
   A. radiant energy from sunlight
   B. geothermal energy from Earth's interior
   C. hydroelectric energy from running water
   D. energy from atomic nuclei splitting apart

9. How are solar energy and wind energy similar?
   A. Both are able to produce electricity.
   B. Both are nonrenewable sources of energy.
   C. Both rely on the power generated by the sun.
   D. Both produce high amounts of carbon dioxide.

10. Falling water runs through a hydroelectric power plant. The falling water turns turbines in the plant to produce electricity. The electricity is transmitted into the homes of people. There, the power is used to operate electric heaters. Which best identifies the energy transformation that has taken place?
    A. kinetic energy → electrical energy → thermal energy
    B. electrical energy → kinetic energy → potential energy
    C. potential energy → thermal energy → electrical energy
    D. thermal energy → electrical energy → chemical energy

11. What causes us to hear sounds?
    A. radiant energy from the sun
    B. acoustic energy from vibrations of air
    C. thermal energy released from molecular motion
    D. chemical energy discharged from photosynthesis

12. Chemical energy is an important source of power.

    In your **Answer Document**, describe chemical energy. Then explain why scientists consider chemical energy to be a form of potential energy. (2 points)

13. In your **Answer Document**, draw a diagram of a baseball rolling down a small hill. On your diagram, identify where the baseball's potential energy is greatest and lowest. Then indicate where the baseball's kinetic energy is the greatest and lowest. (4 points)

14. Renewable energy is an essential source of power today.

    In your **Answer Document**, identify two types of renewable energy sources. Then, give one risk or one benefit of using each type. (4 points)

CHECKLIST OF PHYSICAL SCIENCE BENCHMARKS

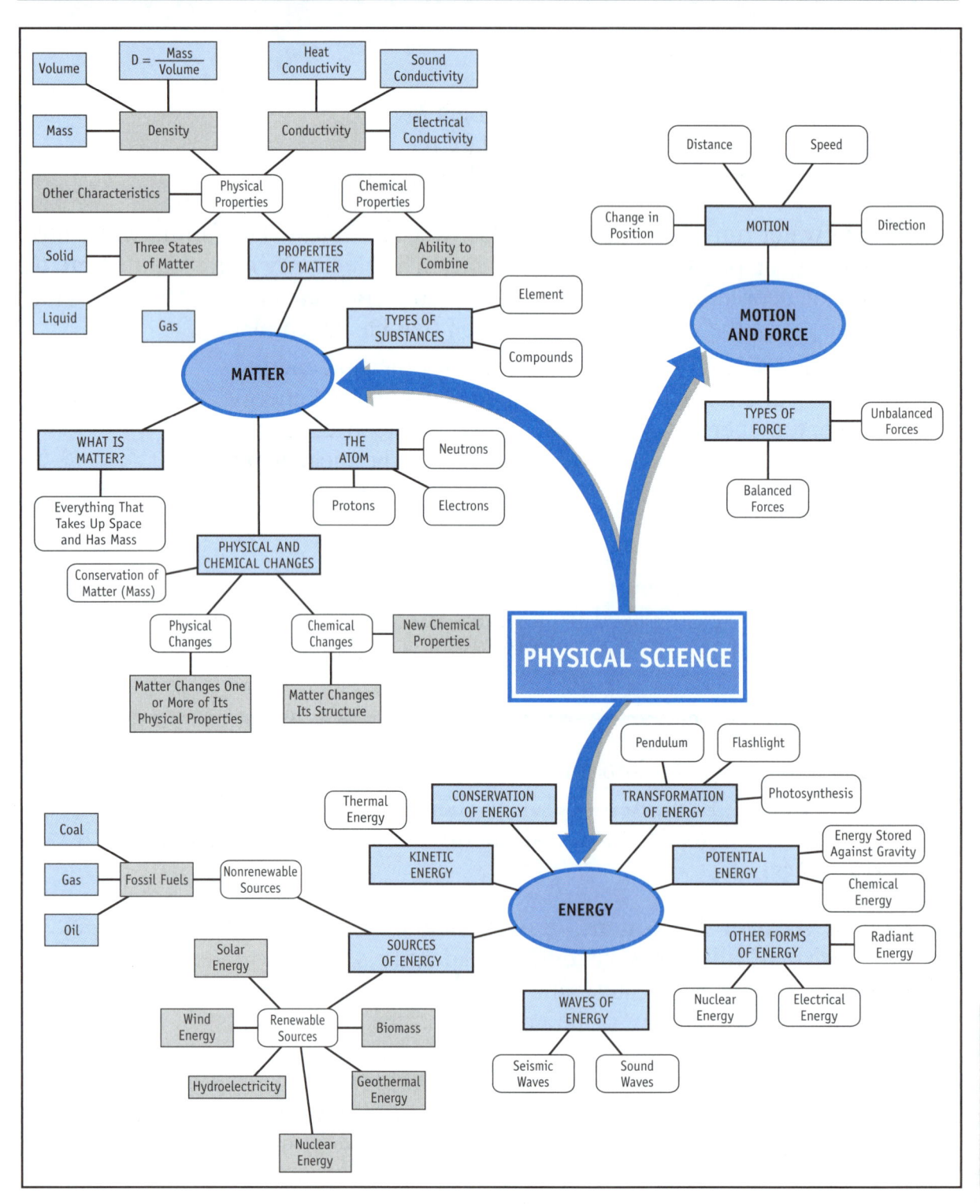

PHYSICAL SCIENCE

MATTER

- **Properties of Matter**
  - Physical Properties
    - Density: $D = \dfrac{\text{Mass}}{\text{Volume}}$
      - Volume
      - Mass
    - Conductivity
      - Heat Conductivity
      - Sound Conductivity
      - Electrical Conductivity
    - Other Characteristics
    - Three States of Matter
      - Solid
      - Liquid
      - Gas
  - Chemical Properties
    - Ability to Combine
- **Types of Substances**
  - Element
  - Compounds
- **What Is Matter?**
  - Everything That Takes Up Space and Has Mass
- **The Atom**
  - Neutrons
  - Protons
  - Electrons
- **Physical and Chemical Changes**
  - Conservation of Matter (Mass)
  - Physical Changes
    - Matter Changes One or More of Its Physical Properties
  - Chemical Changes
    - New Chemical Properties
    - Matter Changes Its Structure

MOTION AND FORCE

- **Motion**
  - Distance
  - Speed
  - Change in Position
  - Direction
- **Types of Force**
  - Unbalanced Forces
  - Balanced Forces

ENERGY

- **Kinetic Energy**
  - Thermal Energy
- **Conservation of Energy**
- **Transformation of Energy**
  - Pendulum
  - Flashlight
  - Photosynthesis
- **Potential Energy**
  - Energy Stored Against Gravity
  - Chemical Energy
- **Other Forms of Energy**
  - Radiant Energy
  - Nuclear Energy
  - Electrical Energy
- **Waves of Energy**
  - Seismic Waves
  - Sound Waves
- **Sources of Energy**
  - Nonrenewable Sources
    - Fossil Fuels
      - Coal
      - Gas
      - Oil
  - Renewable Sources
    - Solar Energy
    - Wind Energy
    - Biomass
    - Hydroelectricity
    - Geothermal Energy
    - Nuclear Energy

TESTING YOUR UNDERSTANDING

1. A black iron fence starts to rust as the iron combines with oxygen in the air. How do scientists know that this is a chemical change?
   A. The iron has melted.
   B. The change is reversible.
   C. The color of the fence has changed.
   D. Rust has its own unique chemical properties.

   PS: A
   G6.2

   A group of bike riders took a 4 hour trip. During the first 3 hours, they traveled a total of 50 kilometers. During the last hour, they traveled only 10 kilometers.

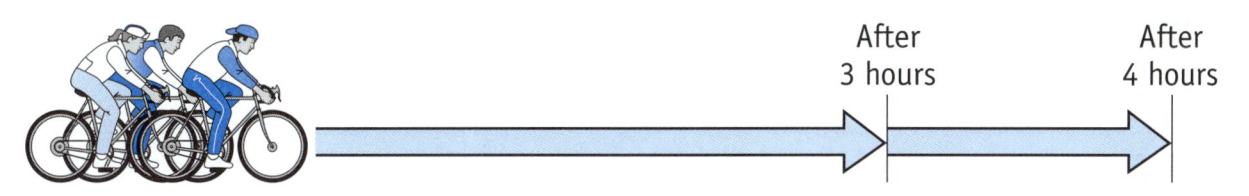

   After 3 hours   After 4 hours

2. What was the total distance traveled by the group during the entire trip?
   A. 15 km   C. 40 km
   B. 30 km   D. 60 km

   PS: B
   G8.2

3. What was the mean speed of the cyclists for all 4 hours?
   A. 10 km / hr   C. 17 km / hr
   B. 15 km / hr   D. 60 km / hr

   PS: B
   G8.2

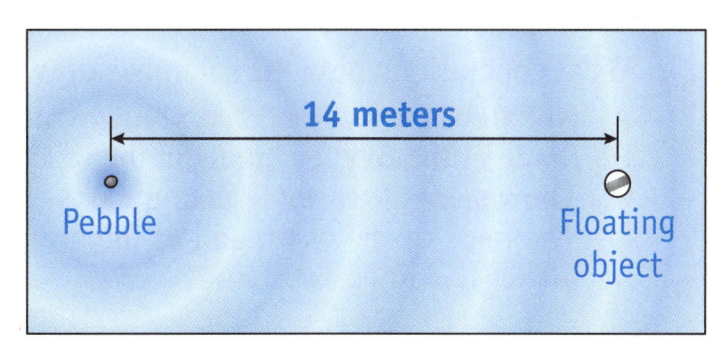

4. A ripple tank is a shallow container of water used to study the properties of waves. A group of students drop a pebble into one side of the ripple tank. They measure the ability of the water to lift a floating object at the opposite end of the tank. What hypothesis are the students testing?

   14 meters
   Pebble   Floating object

   A. Waves can transfer energy.
   B. Sound waves can travel through a body of water.
   C. Balanced forces will not change an object's motion.
   D. Waves travel more slowly through dense materials than through less dense ones.

   PS: D
   G8.4

5. Which appliance converts electrical energy into thermal energy?

A.   B.   C.   D.

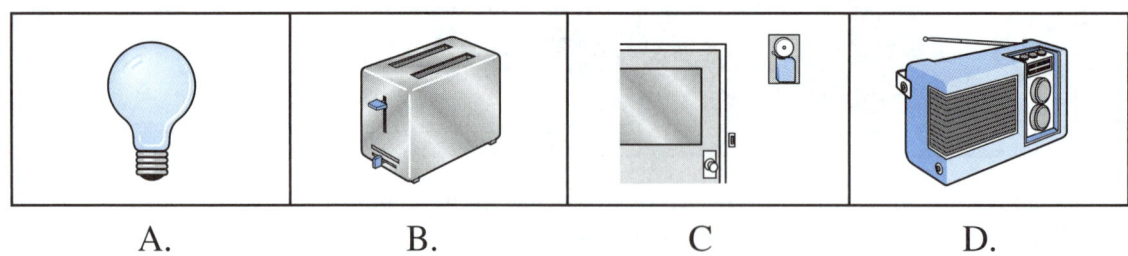

Use the following information to answer questions 6 and 7.

A group of scientists carefully measure quantities of oxygen gas and hydrogen gas in a laboratory. Then they mix these gases. A reaction occurs that produces water.

Element	Before the Reaction	After the reaction
Hydrogen	4 grams	0 grams
Oxygen	32 grams	0 grams
Water	0 grams	36 grams

6. How do the scientists know that a chemical change has occurred?
   A. Water is a liquid, while the other substances are gases.
   B. The water can be separated into hydrogen and oxygen by physical methods.
   C. The chemical properties of the water are identical to those of oxygen and hydrogen.
   D. A new substance has formed with different chemical properties than the original substances.

7. What conclusion could the scientists draw from this experiment?
   A. Not all of the hydrogen combined with the oxygen.
   B. Some of the oxygen atoms in the reaction were lost.
   C. The water is less stable than the oxygen and hydrogen gas.
   D. The mass of the water and of the gases was the same since the amount of matter remained constant.

8. Which practice depletes a nonrenewable source of energy?
   A. harvesting trees in a rainforest
   B. burning coal to generate electricity in a power plant
   C. using steam from below Earth's surface to turn a turbine
   D. building a dam to use falling water to generate electricity

9. What is conserved during a chemical reaction?
   A. energy only
   B. matter only
   C. both matter and energy
   D. neither matter nor energy

   PS: A
   G7.1

10. A student throws a ball onto the ground. At first, the ball rolls rapidly. After a few moments the ball stops moving. Why did the ball stop moving?
    A. The ball was subject to balanced forces.
    B. Motion always requires an unbalanced force to continue.
    C. An unbalanced force, friction, stopped the motion of the ball.
    D. The student did not throw the ball hard enough to keep it rolling.

    PS: B
    G8.3

11. What causes us to hear sounds?
    A. vibrations in the air producing waves
    B. radiant energy from the sun traveling to Earth
    C. stored chemical energy changing into electricity
    D. atoms and molecules moving rapidly in a substance

    PS: D
    G8.5

12. At which location in the diagram does the train have the greatest potential energy?
    A. Location 1
    B. Location 2
    C. Location 3
    D. Location 4

    PS: D
    G7.2

13. Which shows a change from chemical to electrical to radiant energy?
    A. A candle burns and lights up the room.
    B. A display of fireworks lights up the night sky.
    C. A car battery causes a car's headlights to shine.
    D. Rocks roll down a steep mountain in an avalanche.

    PS: D
    G7.5

14. A group of students is determining an object's density. They first measure its mass. What must they next measure to calculate the density of the object?
    A. its area
    B. its temperature
    C. its height
    D. its volume

    PS: A
    G6.1

15. A student watches several cars race around an outdoor track. What information would the student need to describe the motion of the cars?
    A. their shape and mass
    B. their length and height
    C. their speed and direction
    D. their temperature and volume

    PS: B
    G8.2

16. In a chemical reaction, substances change their properties but matter is conserved.

In your **Answer Document**, explain how substances change in a chemical reaction. Then explain how matter in a chemical reaction is conserved. (2 points)

*PS: A
G6.2
G7.1*

17. In today's world, people use many different kinds of energy to meet their everyday needs.

In your **Answer Document**, describe two types of energy that people in Ohio often use in their homes. (2 points)

*PS: C
G7.3*

18. Renewable and nonrenewable energy sources are important to our lives.

In your **Answer Document**, identify one example of each. Then describe two ways that renewable and nonrenewable energy sources can be managed or conserved. (4 points)

*PS: C
G6.8*

19. Chemical and physical changes occur all around us.

In your **Answer Document**, describe two differences between a chemical and a physical change. Then provide one example of each type of change. (4 points)

*PS: A
G6.4*

CHECKLIST OF PHYSICAL SCIENCE BENCHMARKS

Directions. Now that you have completed this unit, place a check (✔) next to those benchmarks you understand. If you are having trouble recalling information about any of these benchmarks, review the lesson indicated in the brackets.

☐ You should be able to relate uses, properties and chemical processes to the behavior and/or arrangement of the small particles that compose matter. [Chapter 7]

☐ You should be able to describe the motion of objects and the effects of forces on an object. [Chapter 8]

☐ You should be able to describe renewable and nonrenewable sources of energy (e.g., solar, wind, fossil fuels, biomass, hydroelectricity, geothermal and nuclear energy) and the management of these sources. [Chapter 9]

☐ You should be able to describe that energy takes many forms: some forms represent kinetic energy and other forms represent potential energy; and that during energy transformations, the total amount of energy remains constant. [Chapter 9]

UNIT 4: LIFE SCIENCES

In this unit, you will learn about the **life sciences** — the study of all living things. You will learn how living things function and how they interact with the physical environment. You will learn about the **cell**, the basic unit of life.

You will also explore tissues and organs, types of reproduction, inherited traits, ecosystems, and how organisms adapt or fail to adapt to environmental changes.

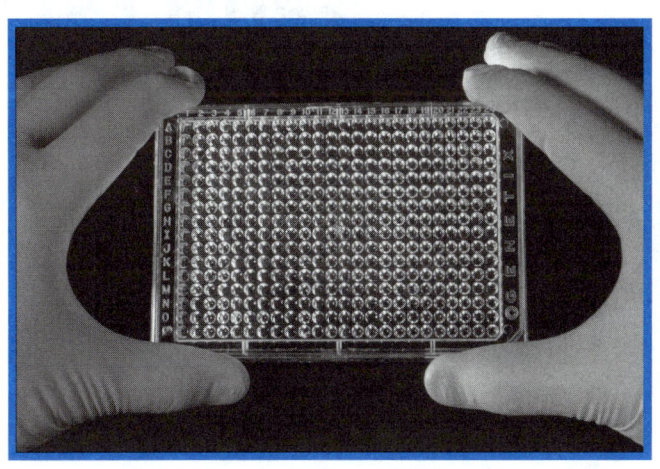

A tray containing a human DNA sequence.

CHAPTER 10: FROM CELLS TO MULTICELLULAR ORGANISMS

In this chapter, you will learn about different types of cells and their parts. You will study cellular processes, including how cells use energy and matter. You will also study tissues, organs, and the body plans of multicellular organisms.

CHAPTER 11: REPRODUCTION AND INHERITED TRAITS

In this chapter, you will learn how some organisms reproduce. You will also learn how each organism can be described as a combination of inherited traits.

CHAPTER 12: ECOSYSTEMS AND ADAPTATION TO ECOLOGICAL CHANGE

In this chapter, you will learn about various types of ecosystems, how energy and matter flow through an ecosystem, and how ecosystems undergo change. You will also learn how organisms either adapt to environmental change or become extinct.

CHAPTER 10

FROM CELLS TO MULTICELLULAR ORGANISMS

In this chapter, you will learn about organisms. An **organism** is any living thing that can live on its own. You will also learn how all organisms are made of cells.

— MAJOR IDEAS —

- **A.** The **cell** is the basic unit of all living things. All cells carry on functions to live, grow, and reproduce.
- **B.** Plant cells change the energy of sunlight into chemical energy through **photosynthesis**. **Cellular respiration** releases this energy.
- **C.** A **tissue** has similar cells that work together in a multicellular organism. An **organ** has tissues that work together to perform one or more functions. An **organ system** has organs that work together for a specific function.
- **D.** Multicellular organisms have various **body plans** and **internal structures**.

WHAT IS A CELL?

If you looked at a piece of your skin under a microscope, you would find that it is composed of tiny, bubble-like shapes known as **cells**. In fact, all living things are made up of cells. All cells share certain common characteristics:

★ They are surrounded by a **cell membrane**. The cell membrane holds the cell together. Its function is to control what enters and leaves the cell.

★ Cells are mainly composed of **fluid** (cytoplasm). This fluid is made up of chemicals and special structures that help the cells to live, grow, and reproduce.

★ Each cell has hereditary material (**DNA**). This DNA was received from a pre-existing cell. DNA provides the instructions for how each cell operates.

CELLULAR FUNCTIONS AND STRUCTURES

Many of the basic functions of organisms are carried out by individual cells.

TRANSPORT

An individual cell controls what enters and leaves its cell membrane. It can allow oxygen, water, or food to enter, and can send out waste from the cell.

EQUILIBRIUM

All cells must maintain **equilibrium** — stable internal conditions. For example, a cell must control how much water it has and its temperature.

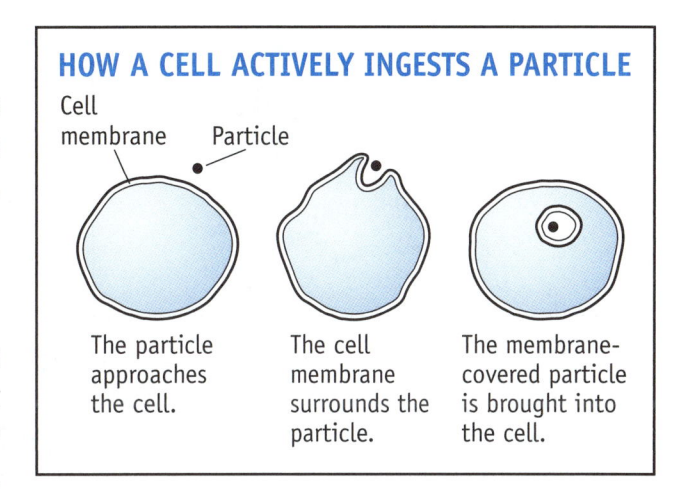

HOW A CELL ACTIVELY INGESTS A PARTICLE

The particle approaches the cell.

The cell membrane surrounds the particle.

The membrane-covered particle is brought into the cell.

GROWTH

All cells can build new molecules, such as protein molecules, based on instructions in their DNA. This capability allows them to grow.

REPRODUCTION

All cells can reproduce themselves. Many cells reproduce simply by copying their DNA and dividing in two. Other cells divide their DNA and combine with other cells to reproduce. You will learn more about reproduction in the next chapter.

APPLYING WHAT YOU HAVE LEARNED

◆ A scientist once said, in describing a cell — "one name, many functions." What did the scientist mean by this? _____
_____

CELL STRUCTURES

To perform their functions, different types of cells have special structures. Bacteria, for example, are the simplest type of cells. They have a cell wall and a cell membrane covering the cell. The DNA of bacteria is just bunched up like a curled rubber band inside the cell.

A BACTERIA CELL

- Cytoplasm
- Cell wall
- DNA
- Cell membrane
- Ribosomes
- Flagellum

AN ANIMAL CELL

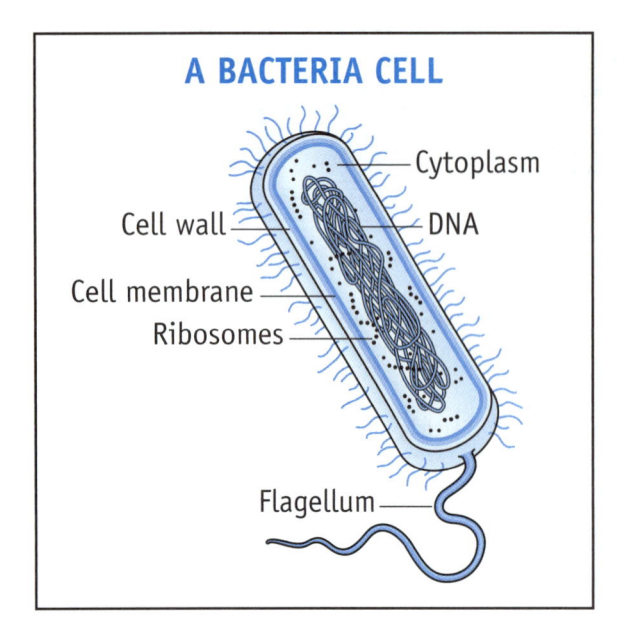

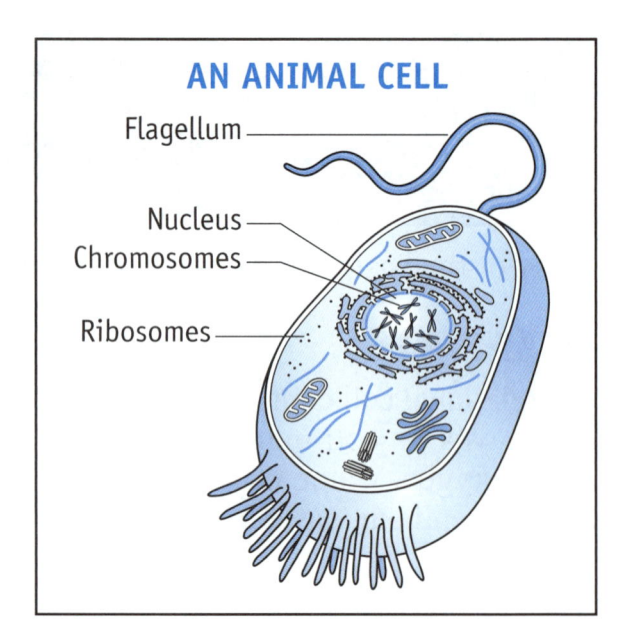

- Flagellum
- Nucleus
- Chromosomes
- Ribosomes

Other cells are more complex. A large nucleus inside the cell holds the cell's DNA. The nucleus is covered by its own double membrane. Other special structures exist to help the cell perform its essential functions. Some of these structures help move food, gases, water and other particles around the cell. Other structures help the cell break down sugars to release energy in a process known as **cellular respiration**. Still other structures help the cell make new molecules of its own.

PLANT CELLS

Although both plants and animals are made up of cells, they are not the same. **Plant cells have several special additional structures that make them unique.**

CELL WALLS

The cell membranes of plants are covered by a rigid, outer cell wall that supports and protects the plant. A plant's cell walls have pores that allow it to interact with the environment. Because of its cell walls, a plant cell generally does not move. In contrast, an animal cell has no cell walls.

VACUOLES

These large, fluid-filled "bubbles" within plant cells hold water, enzymes and waste. Vacuoles can be quite large, taking up to 90 percent of a plant cell's volume.

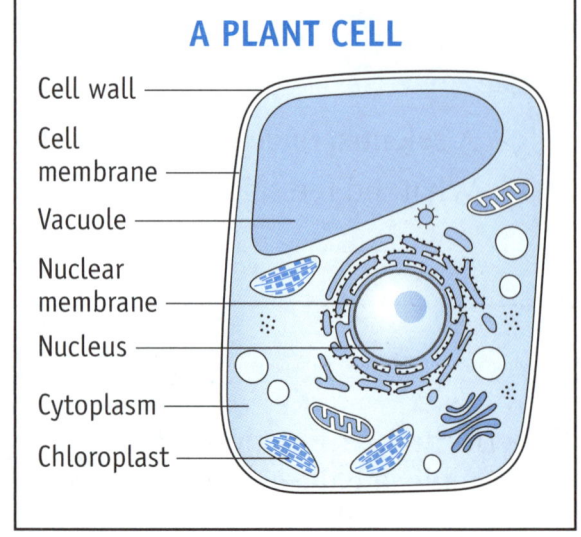

A PLANT CELL

- Cell wall
- Cell membrane
- Vacuole
- Nuclear membrane
- Nucleus
- Cytoplasm
- Chloroplast

A typical plant cell.

CHAPTER 10: FROM CELLS TO MULTICELLULAR ORGANISMS

CHLOROPLASTS

Plant cells have these special structures, which allow plants to conduct **photosynthesis** — the conversion of sunlight into chemical energy. Chloroplasts are green, giving plants their distinctive color.

HOW ANIMAL CELLS DIFFER FROM PLANT CELLS

Animal Cells	Plant Cells
• Can move about	• Are confined by cell walls
• Must consume other organisms for food	• Conduct photosynthesis
• Expel or dissolve wastes	• Have large vacuoles for enzymes and waste

APPLYING WHAT YOU HAVE LEARNED

◆ On a separate sheet of paper, create your own drawing of a typical animal cell and plant cell. Be sure to label their main parts, indicating the main function of each part.

HOW CELLS GAIN AND RELEASE ENERGY

All forms of life require energy. Plants obtain their energy from the sun. Other organisms acquire energy by eating plants or other organisms.

PHOTOSYNTHESIS

Plants obtain energy from radiant sunlight and change it to chemical energy. Energy from the sun travels to Earth's surface. A green plant needs only a few seconds to capture the energy in sunlight and store it in the form of a chemical bond. Plants use the energy from sunlight to create a form of sugar out of carbon dioxide and water. Plants release oxygen as a by-product.

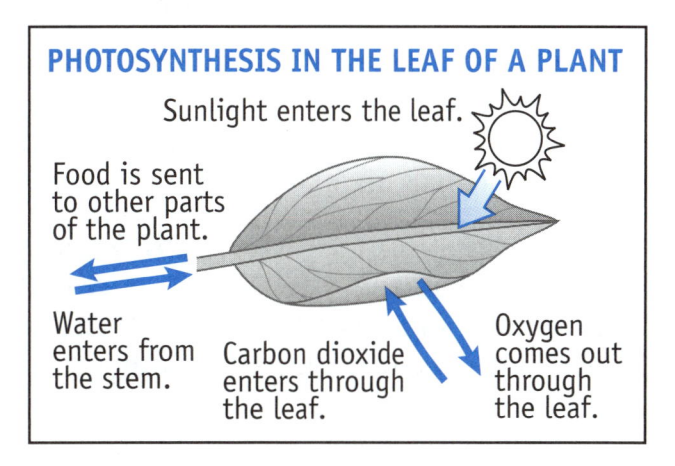

PHOTOSYNTHESIS IN THE LEAF OF A PLANT

Sunlight enters the leaf.
Food is sent to other parts of the plant.
Water enters from the stem.
Carbon dioxide enters through the leaf.
Oxygen comes out through the leaf.

The process of converting light energy into stored chemical energy is called **photosynthesis**:

> **Carbon dioxide + water + sunlight → glucose + oxygen (sugar)**

CELLULAR RESPIRATION

During **cellular respiration**, cells break down glucose to release energy. Cellular respiration is not the same as breathing, but there are similarities: both require oxygen and produce carbon dioxide. Cells use oxygen in a series of reactions to change glucose back into carbon dioxide and water. These reactions release the energy stored in the sugar by photosynthesis. In fact, cellular respiration is almost the reverse of photosynthesis.

APPLYING WHAT YOU HAVE LEARNED

◆ Make a chart comparing *photosynthesis* and *cellular respiration*.

	Photosynthesis	Cellular Respiration
What occurs?		
How are they different?		

MULTICELLULAR ORGANISMS

Many cells are **unicellular organisms**. These are single cells that can live on their own. Bacteria are unicellular **organisms**. So are some animal-like cells, such as an amoeba.

Other cells form **multicellular organisms**. These are organisms with two or more cells. In multicellular organisms, not all cells are alike. Each cell has its own specialized functions.

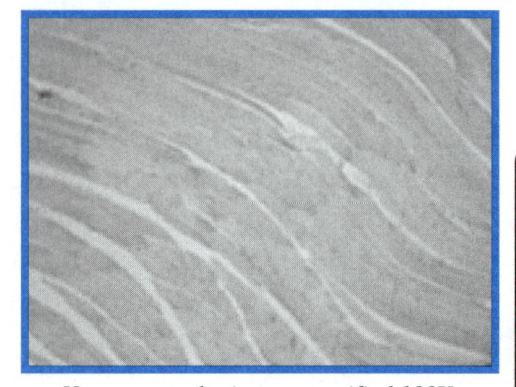

Human muscle tissue magnified 100X.

CHAPTER 10: FROM CELLS TO MULTICELLULAR ORGANISMS 133

TISSUES

A **tissue** is a group of cells with similar characteristics that work together. For example, muscle tissue in animals is made up of similar muscle cells. They act together to form a muscle. Other types of tissue in many animals include **nervous tissue** and **skin tissue**.

ORGANS

An **organ** is made up of two or more tissues working together. Each organ generally performs a specialized function. For example, the heart is an important organ in the body. The **heart** has the function of supplying the body with blood. The heart's contractions pump blood throughout the body. Every day, the average human heart beats about 100,000 times.

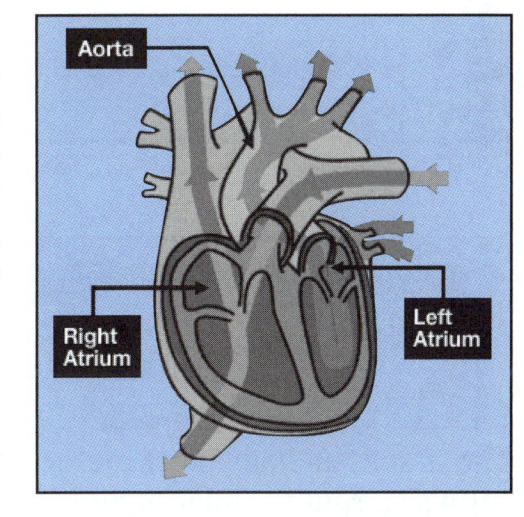

ORGAN SYSTEM

An **organ system** is a group of organs that work together in an organism to perform a specific function. For example, in humans and other animals the heart works together with arteries, veins, and capillaries to carry blood throughout the body. Together, the heart and these blood vessels make up the **circulatory system**.

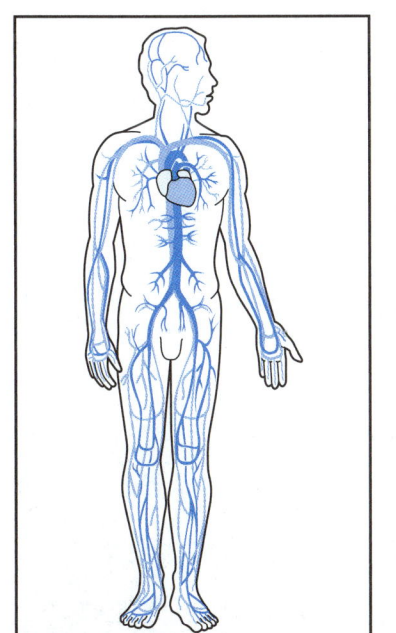

The circulatory system delivers a continuous flow of blood to the body of an animal, supplying its cells with oxygen and nutrients, and carrying away wastes. **Arteries** carry oxygen-rich blood from the heart to the rest of the body. **Veins** return blood to the heart. The **capillaries** are very fine blood vessels between the arteries and veins. They allow the blood to provide oxygen to neighboring cells in exchange for carbon dioxide.

APPLYING WHAT YOU HAVE LEARNED

◆ On page 30 of this book, you were asked to start your own science glossary. On the next page is a list of important terms in this chapter. Add each of these to your glossary with its definition.

- ✦ Cell: _____
- ✦ Unicellular organism: _____
- ✦ Multicellular organism: _____
- ✦ Tissue: _____
- ✦ Organ: _____
- ✦ Organ system: _____

TYPES OF MULTICELLULAR ORGANIZATION

The tissues and organs of multicellular organisms can be arranged in many different ways. Multicellular organisms have a variety of body plans and internal structures. For example, the body of a multicellular organism might be **symmetrical** (*the same on each side*), **asymmetrical** (*dissimilar on each side*) or **radial** (*spread out, like the branches of a tree*).

PLANT BODY PLANS AND STRUCTURES

Plants generally organize their bodies into several key organs. They do not move from place to place, but they can make their own food.

ROOTS

Roots anchor the plant in the ground. They prevent gravity or the wind from tipping the plant over. A second function of the roots is to absorb water and minerals from the soil. Roots often spread out in a *radial pattern*.

STEMS

Stems hold the leaves of the plant. They connect the leaves to the roots, and carry food and water back and forth between these parts. Stems sometimes spread out in a radial pattern, like the branches of a tree. Other stems just go straight up.

Plant roots absorb water and nutrients.

LEAVES

Leaves make food for a plant through the process of photosynthesis. Tiny pores in the leaves absorb carbon dioxide from the air and expel oxygen. The shape and position of a leaf allows it to absorb the maximum amount of sunlight.

Mosses are plants that grow close to the ground.

Although most plants share these structures, they have many different body plans and specific structures. For example, some stems are very tube-like, like a stalk of celery. Other plants, like mosses, lack these tiny tubes. Some plants have flowers to reproduce, while other plants do not.

APPLYING WHAT YOU HAVE LEARNED

As you have just learned, all plants have roots, stems, and leaves, but these can be organized in different ways. Name two types of plants and describe the characteristics of each one. You may want to consult an encyclopedia or the Internet to find this information.

Type of Plant	Description of its roots	Description of its stem	Description of its leaves
1.			
2.			

ANIMAL BODY PLANS AND STRUCTURES

Unlike plants, animals have the ability to move. However, they cannot make their own food as a plant can. Therefore, they must ingest food by eating plants or other animals. There are a great variety of body types and internal structures for animals, just as there are for plants.

BODY PLANS

Scientists generally divide the body plans of animals into two types — invertebrates and vertebrates.

★ **Invertebrates** have no backbone. For example, jellyfish, lobsters, and insects are all invertebrates. Most of these animals have a shell or other external skeleton. They make up the vast majority of animals on Earth.

★ **Vertebrates** are animals with a backbone. The backbone supports the entire organism, and helps the animal hold and protect its most important internal organs. Birds, fish, reptiles, and mammals are vertebrates since they all have backbones. Vertebrates are generally symmetrical: the left and right side of the body is almost identical. For example, humans have two eyes, two ears, two arms, and two legs.

APPLYING WHAT YOU HAVE LEARNED

Indicate with a check (✔) if each animal has a backbone or not. If you are not sure of the answer, consult an encyclopedia or the Internet.

Animal	Vertebrate	Invertebrate	Animal	Vertebrate	Invertebrate
Shark			Ostrich		
Jellyfish			Butterfly		
Mole			Octopus		

INTERNAL STRUCTURES

The internal structures of animals are generally organized into organ systems — groups of organs working together. For example, whales live in ocean waters. Although air-breathing mammals, they spend about 85 percent of their time submerged. Light does not penetrate water well, so whales have highly developed auditory systems to hear sound and to sense their environment. They have no external ears, but have ears buried inside their heads. Nerves carry signals from their ears to the brain. When a whale swallows food, it travels through the whale's esophagus to a multi-chambered stomach where the food is crushed. They also have highly developed respiratory systems, since they can hold their breath for long periods under water.

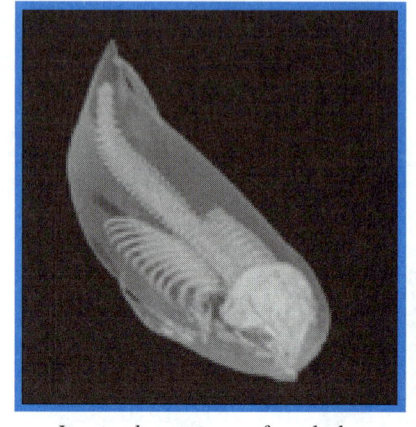

Internal structure of a whale.

CHAPTER 10: FROM CELLS TO MULTICELLULAR ORGANISMS   137

APPLYING WHAT YOU HAVE LEARNED

✦ To meet all of its needs, a typical vertebrate organizes its body into several organ sytems. Complete the chart below, summarizing each of these systems.

System	Main Function	Organs
Skeletal	Supports the body and helps it to move around.	
Circulatory		
Respiratory	Adds oxygen to the blood; allows the body to expel carbon dioxide.	
Digestive		
Excretory	Removes wastes from the body.	
Nervous	Allows the organism to feel or sense things; controls muscular movements.	
Reproductive	Allows the organism to reproduce.	
Muscular	Allows body parts, such as the arms, to move.	

✦ List four organisms. Investigate each organism you select by looking in an encyclopedia, your school library, or on the Internet. Then complete the chart below.

Organism	Plant or Animal?	Body Plan	Internal Structures
1.			
2.			
3.			
4.			

APPLYING WHAT YOU HAVE LEARNED

◆ Now take two of the organisms from the chart on page 137. Consider how they are similar and how they are different. Then complete the Venn diagram below:

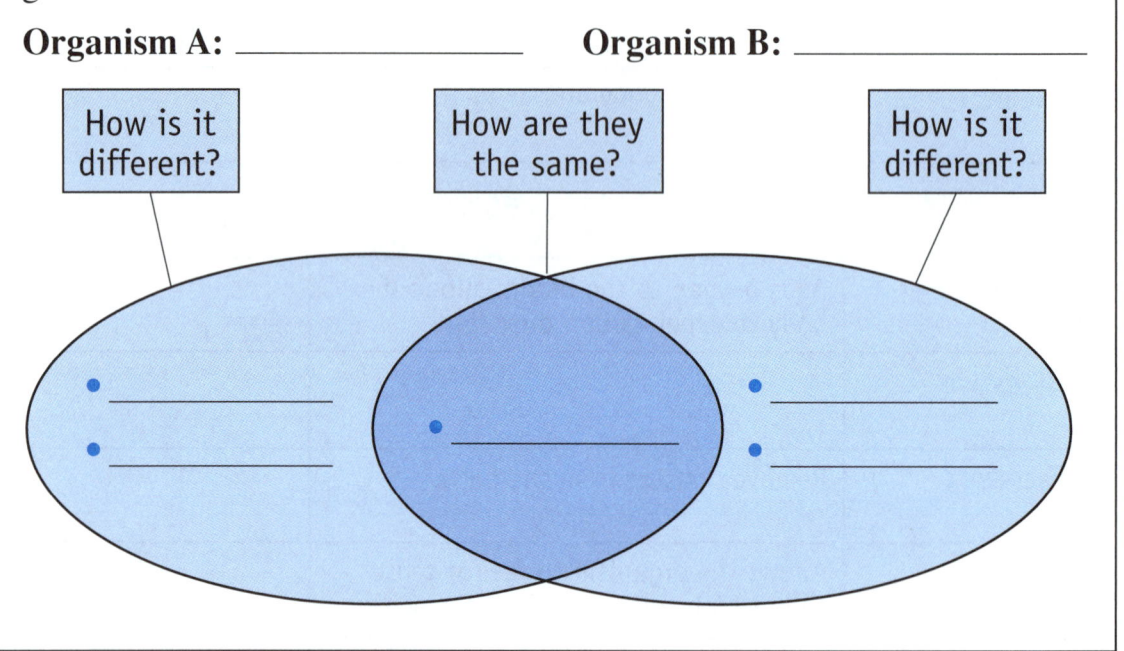

Organism A: _____ **Organism B:** _____

How is it different? How are they the same? How is it different?

- _____
- _____

- _____

- _____

WHAT YOU SHOULD KNOW

☐ You should know that the **cell** is the basic unit of all living things. All cells carry on functions to live, grow, and reproduce.

☐ You should know that **plant** and **animal cells** differ. Plant cells have cell walls, vacuoles, and chloroplasts.

☐ You should know that plant cells change the energy of sunlight into chemical energy through **photosynthesis. Cellular respiration** releases this energy so that the cell can use it for cellular processes.

☐ You should know that a **tissue** is a group of similar cells that work together in a multicellular organism. An **organ** is a group of tissues that work together to perform a specific function. An **organ system** is a group of organs that work together to perform a specific function in an organism.

☐ You should know that multicellular organisms have different **body plans** and **internal structures**.

CHAPTER STUDY CARDS

The Cell

★ The cell is the basic unit of all living things.

★ **Membranes and DNA.** All cells have cell membranes and contain a hereditary blueprint for their structure and processes, known as DNA.

★ **Carry Out Functions of Life.** A cell carries out all the basic functions necessary for life, such as using energy, eliminating waste, maintaining equilibrium, growing and reproducing.

Types of Cells

★ **Animal Cells.** The cells of animals can move around, but they lack chloroplasts and cannot make food.

★ **Plant Cells.** The cells of plants contain cell walls, chloroplasts, vacuoles; plant cells cannot move but can make their own food.

Types of Organisms

★ **Unicellular.** Organism with only one cell.

★ **Multicellular.** An organism that has several cells with specialized functions.

Multicellular Organisms

★ **Tissue.** Tissue is made up of similar cells that work together. For example, muscle tissue.

★ **Organ.** An organ is made up of two or more tissues that work together. For example, the heart, liver, and lungs in humans and many other animals.

★ **Organ System.** An organ system is a group of organs that work together to carry out some function. For example, circulation, respiration, and digestion.

Body Plans and Structures of Multicellular Organisms

★ Body plans and internal structures differ from one multicellular organism to another.

★ **Plants.** Plants have roots, stems, and leaves; plants differ as to whether their stems have tubes, whether they have flowers to reproduce, etc.

★ **Animals.** There are two main plans:
  • **Invertebrates.** Have no backbone.
  • **Vertebrates.** Have a backbone.
  • Their internal structures are organized into different organ systems.

CHECKING YOUR UNDERSTANDING

1. Which is the correct order of organization from smallest to largest?
   A. cell → organ → tissue → organ system
   B. tissue → cell → organ system → organ
   C. tissue → cell → organ → organ system
   D. cell → tissue → organ → organ system

LS: A G6.2

♦ Examine the Question
♦ Recall What You Know
♦ Apply What You Know

> **HINT**
> To answer this question correctly, you must understand how complex multicellular organisms are organized. The cell is the simplest unit of life, found in all organisms. Several cells form tissues; tissues form organs, and organs form organ systems. Only **Choice D** correctly shows this sequence, from the simplest to the most complex.

Now try answering some additional questions on your own:

2. The diagram to the right represents a process that occurs in the cells of a plant. What process is illustrated?
   A. cell division
   B. reproduction
   C. photosynthesis
   D. cellular respiration

   LS: A
   G6.3

3. There are several differences between plant and animal cells. What is one structure of plant cells that animal cells lack?
   A. nucleus
   B. chloroplasts
   C. cell membrane
   D. DNA

   LS: A
   G6.3

4. Which of these groups contains only animals that are vertebrates?
   A. bear, lion, whale
   B. dog, bear, octopus
   C. earthworm, crayfish, snail
   D. lobster, earthworm, goldfish

   LS: A
   G7.1

5. Plants can survive in a clear, closed container with water and nutrients but without animals. Animals cannot survive in a closed container without plants. Why are animals unable to survive without plants?
   A. Plant and animal cells need water to survive.
   B. Plants cannot move, while animals can roam around.
   C. Plant cells can create their own food, but animals cannot.
   D. Plants take in and give off water; animals only take in water.

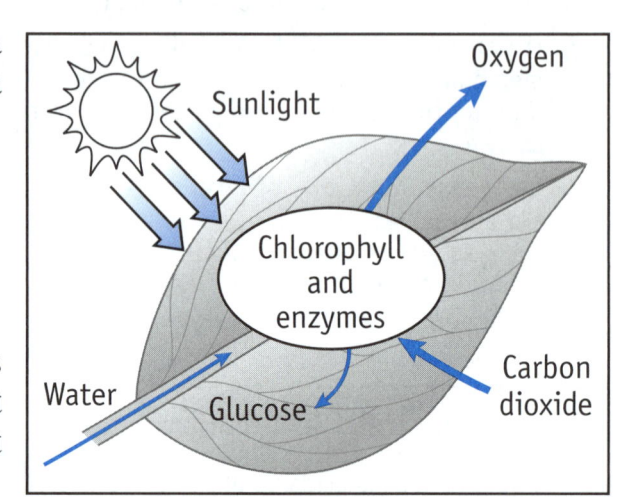

   LS: A
   G6.3

CHAPTER 10: FROM CELLS TO MULTICELLULAR ORGANISMS   141

6. A student is creating a diagram to show how photosynthesis works. In addition to the plant's leaves, what would be essential to include in the diagram?
    A. the bark since it protects the stem
    B. the roots because they bring in water
    C. the soil since it holds the plant in the ground
    D. bacteria since they provide nitrogen to the plant

    LS: A
    G7.1

7. In a multicellular organism, such as a spider, which of these is made up of all the others?
    A. cells
    B. tissues
    C. organs
    D. organ systems

    LS: A
    G6.2

8. What cellular process converts energy from sunlight into chemical energy that organisms can use?
    A. respiration, which uses oxygen
    B. digestion, which breaks down nutrients
    C. fermentation, which breaks down sugars
    D. photosynthesis, which produces a form of sugar

    LS: A
    G6.3

9. In what process does a cell break down sugars to obtain energy for its functions?
    A. digestion   C. respiration
    B. circulation   D. transportation

    LS: A
    G5.1

10. What does the illustration to the right show?
    A. a cell membrane
    B. a piece of tissue
    C. an organ system
    D. a unicellular organism

    LS: B
    G6.2

11. There are many ways in which plant and animal cells are similar and some ways in which they are different.

    In your **Answer Document**, create a Venn diagram comparing the characteristics of plant and animal cells.

    In your diagram, show two similarities of plant and animal cells and two differences between them. (4 points)

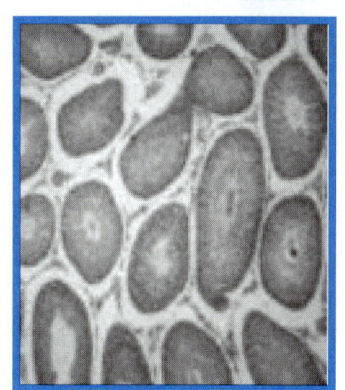

    LS: A
    G6.3

Use the following diagram to answer questions 12–16

| Maple Tree | Jellyfish | Amoeba | Eagle |

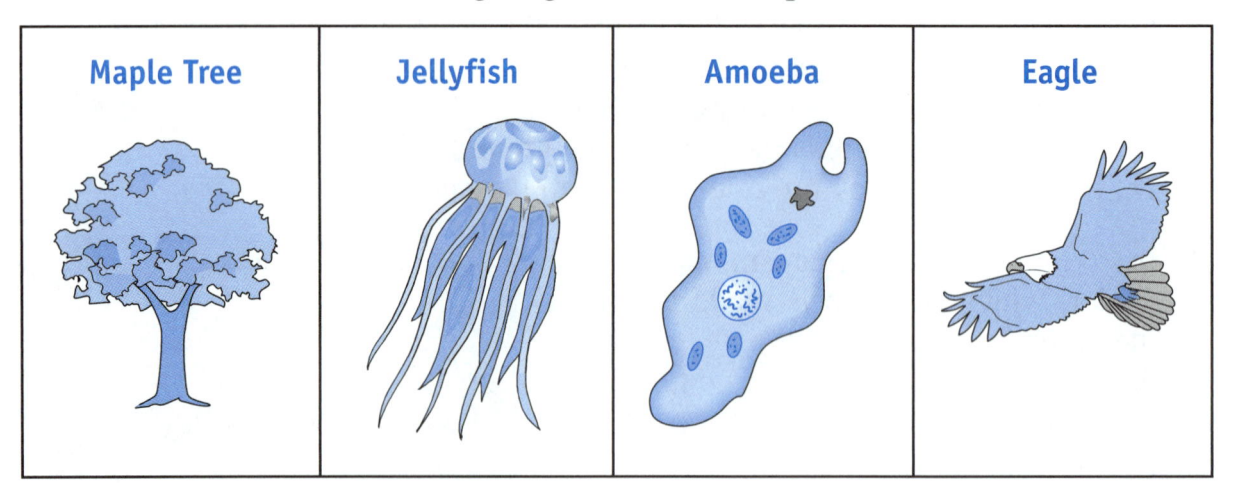

12. Which of these organisms has a symmetrical body plan?
    A. maple tree
    B. amoeba
    C. jellyfish
    D. eagle

    LS: A
    G7.1

13. Which of these organisms is a unicellular organism?
    A. maple tree
    B. amoeba
    C. jellyfish
    D. eagle

    LS: A
    G6.2

14. Which of these organisms is a multicellular invertebrate?
    A. maple tree
    B. amoeba
    C. jellyfish
    D. eagle

    ♦ Examine the Question
    ♦ Recall What You Know
    ♦ Apply What You Know

    LS: A
    G7.1

15. Which of these organisms has a radial body plan?
    A. maple tree
    B. amoeba
    C. jellyfish
    D. eagle

    LS: A
    G7.1

16. Multicellular organisms have a wide variety of body plans and internal structures.

    In your **Answer Document**, select two multicellular organisms from the diagram above.

    Then make a Venn diagram showing two similarities and two differences in the body plans or internal structures of the organisms you have chosen. (4 points)

    LS: A
    G7.1

CHAPTER 11

REPRODUCTION AND INHERITED TRAITS

In this chapter, you will learn how living organisms reproduce themselves. Reproduction is essential to the continuation of each **species**, or type of organism. You will also learn how every living organism can be seen as a combination of inherited traits.

— MAJOR IDEAS —

A. Some characteristics, called **traits**, are inherited from an organism's parent or parents. Other behaviors may be learned.

B. Individual organisms do not live forever. Because they reproduce, the **species** continues and **inherited traits** are passed from one generation to the next.

C. In **asexual reproduction**, all inherited traits come from a single parent. Asexual reproduction limits the spread of new characteristics.

D. In **sexual reproduction**, an egg and sperm unite and some traits come from each parent. The offspring is never identical to both parents. New combinations of traits may increase or decrease an organism's chances for survival.

INHERITED TRAITS

All **organisms** have distinct characteristics. These characteristics include such things as size, shape, and habits. Some characteristics develop in response to the environment.

For example, you may get sunburned from being in the sun. Or you may learn some behaviors while interacting with the environment: you may learn to like pancakes with maple syrup, or to be polite to others.

Inherited Traits. Learned behaviors can and often do change. For instance, you may get tired of eating pancakes for breakfast in the morning. Other characteristics, or **traits**, are inherited. These traits cannot be changed. For example, your height was inherited from your parents. So, too, was the color of your eyes. No matter how hard you try, you cannot change your height or the color of your eyes.

APPLYING WHAT YOU HAVE LEARNED

Some people show dimples in their cheeks when they smile. Others, no matter how hard they try, cannot do this. When people put their hands together and interlock their fingers, they place either their left or right thumb on top. Other people can curl the sides of their tongue. Some people are able to make a "Vulcan" hand sign like Mr. Spock does on Star Trek. Some people have a regular thumb, while others have a "Hitchhiker's thumb," which can be bent farther backwards. These are all inherited traits. A person who cannot curl his or her tongue cannot learn to do so.

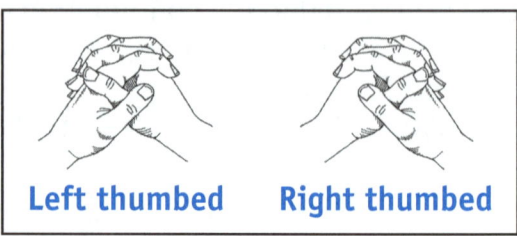

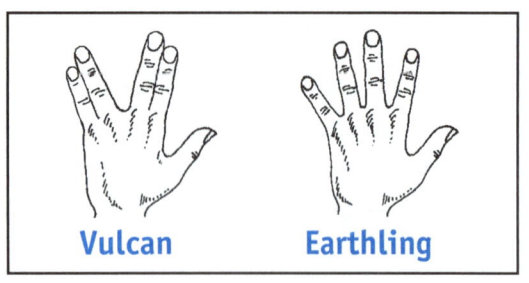

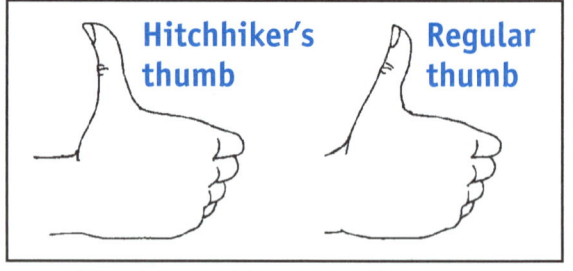

Pick a classmate and make a list of these easily observable traits. Compare your traits with those of your partner.

Characteristic	Your Trait	Classmate's Trait
1. Dimples when smiling	☐	☐
2. Right thumb on top	☐	☐
3. Able to curl tongue	☐	☐
4. "Vulcan" hand sign	☐	☐
5. "Hitchhiker's" Thumb	☐	☐

Every living organism has inherited traits, just as you do. In fact, every living organism can be viewed as a combination of its inherited traits. Inherited traits include more than physical characteristics. Some behaviors, known as **instinctive behaviors**, are also inherited. For example, some species of birds instinctively fly south in the winter. Bats instinctively sleep upside down.

Genes. Inherited traits are controlled by **genes**. Genes are groups of DNA molecules that are inherited from an organism's parent or parents. As you will learn later in this chapter, each inherited trait is controlled by either one or two genes, depending on the type of organism and how it reproduces. Humans inherit one gene for each trait from their mother and a second one for the same trait from their father.

REPRODUCTION

No organism lives forever. Even the oldest tree will eventually die one day. In order for each **species**, or type of organism, to continue, its members must reproduce. This is just as true for simple, unicellular organisms as it is for the most complex, multicellular organisms.

APPLYING WHAT YOU HAVE LEARNED

✦ Why is it important for individuals in a species to reproduce? _____
_____
_____

ASEXUAL REPRODUCTION

Many organisms reproduce by copying themselves. In this case, the traits of the offspring are exactly the same as the parent. This is because the offspring has identical **DNA** — the molecules that provide the "blueprint" for the structure and processes of that organism. There are many different types of asexual reproduction:

★ Unicellular organisms, like bacteria, simply copy their DNA. Then they split into two identical halves. Each new organism has the exact same DNA and inherited traits as the parent cell.

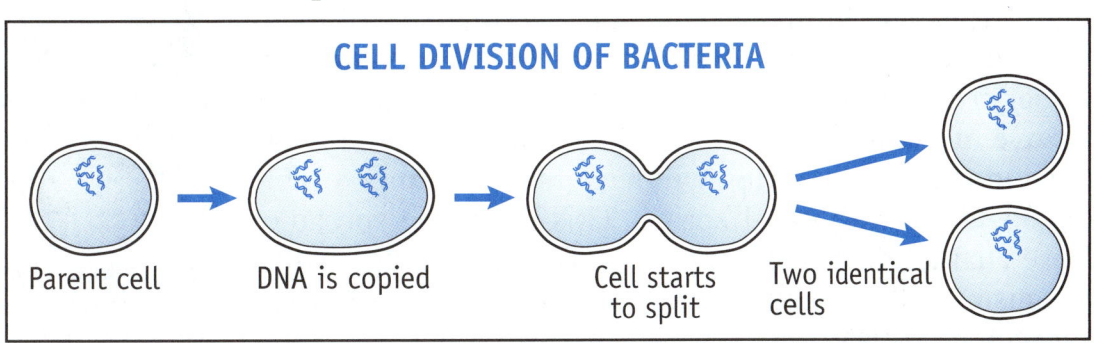

CELL DIVISION OF BACTERIA

Parent cell → DNA is copied → Cell starts to split → Two identical cells

★ New plants can be produced from certain sections of an existing plant. For example, if you bury a carrot root, a carrot plant will grow. This new plant has the same genes as its parent. It has the exact same inherited traits.

★ Fragments of simple animals, like earthworms, can also reproduce a new organism identical to its parent. For example, if certain sections of an earthworm are cut off, they can regenerate the entire organism.

A group of earthworms.

APPLYING WHAT YOU HAVE LEARNED

◆ What is asexual reproduction? _____
_____

◆ Identify two examples of asexual reproduction.
(1) _____   (2) _____

THE CONSEQUENCES OF ASEXUAL REPRODUCTION

ADVANTAGES

In **asexual reproduction**, all inherited traits come from a single parent. This has the advantage of preserving **genetic continuity**. The same genes are reproduced again and again. This also limits the spread of detrimental, or unfavorable, characteristics in a species. However, if an individual in the species has some unfavorable traits, it may not survive in its environment. It will die without reproducing, or its immediate offspring will have the same unfavorable traits and die without offspring. This limits the spread of these characteristics any further. A final advantage is that organisms with asexual reproduction can just multiply without seeking a mate.

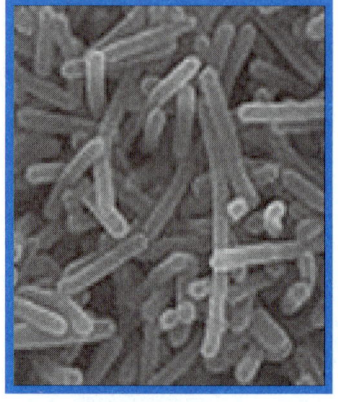

The main advantage of asexual reproduction is that it can be extremely rapid. Bacteria can double their numbers in minutes.

DISADVANTAGES

At the same time, new combinations of traits rarely occur with asexual reproduction. The same genes continually reproduce themselves. If the environment changes, the species may find it very difficult to adapt to the new conditions.

> **APPLYING WHAT YOU HAVE LEARNED**
>
> ◆ List one advantage and one disadvantage of asexual reproduction:
>
> (1) Advantage: _____
>
> _____
>
> (2) Disadvantage: _____
>
> _____

SEXUAL REPRODUCTION

In **sexual reproduction**, two parents are required to produce each offspring. The new offspring inherits some of its traits from each parent. The offspring is therefore never identical to either one of its parents. How does this happen? In species with sexual reproduction, genes come in pairs. Each pair of genes controls one trait. The new organism inherits one gene from each parent.

HOW SEXUAL REPRODUCTION WORKS

These organisms can inherit one gene from each parent because these species have special sex cells. These **sex cells** are used to pass traits on to offspring.

When sex cells divide, each new cell receives only half of the genes of its parent. Some of these sex cells are called **eggs**. Only female members of the species can produce these eggs. Other organisms from the same species, known as males, produce sex cells known as **sperm**. When a sperm from a male organism joins an egg from a female organism of the same species, an offspring is created. The offspring inherits one gene for each trait from each parent.

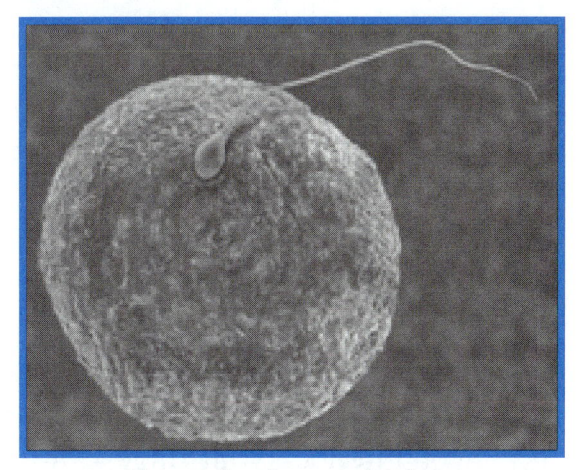

A human female egg as it is being fertilized by a male sperm.

DOMINANT AND RECESSIVE TRAITS

In sexual reproduction, a combination of genes determines whether or not an individual possesses a certain trait. Some traits are **dominant**, while other are **recessive**. Each parent has two genes for each trait, but gives only one of these to its offspring.

★ **Dominant Traits.** If an offspring inherits a dominant trait from either parent, it will show this characteristic. For example, brown eyes are dominant over blue eyes. If a child inherits the gene for blue (b) eyes from his father and the gene for brown eyes (B) from his mother, the child will have brown eyes, since brown eyes are dominant.

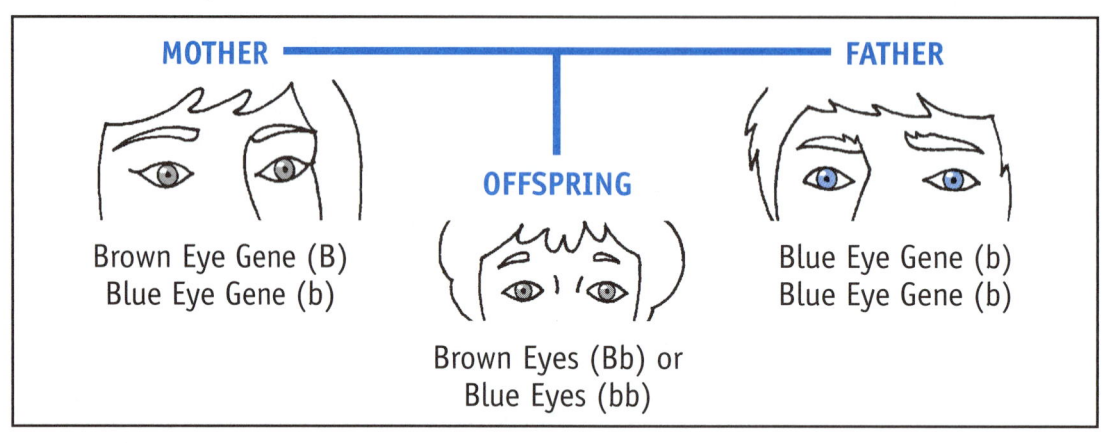

MOTHER — **FATHER**

Brown Eye Gene (B)
Blue Eye Gene (b)

OFFSPRING

Brown Eyes (Bb) or
Blue Eyes (bb)

Blue Eye Gene (b)
Blue Eye Gene (b)

★ **Recessive Traits.** Traits that are not dominant are known as **recessive**. In order for a recessive trait to appear, the offspring must inherit that trait from both parents. In the example above, the child will only have blue eyes if the gene for blue eyes (b) comes from both parents.

APPLYING WHAT YOU HAVE LEARNED

◆ Scientists have found that having unattached earlobes is a dominant trait, while having earlobes that are attached is a recessive trait.
What genes did each person on the right inherit? _____

◆ What genes could the person on the left have inherited? _____
_____

◆ If your left thumb is on top when you interlock your fingers, it is a dominant trait. If your right thumb is on top, you are recessive.
Are you dominant or recessive?

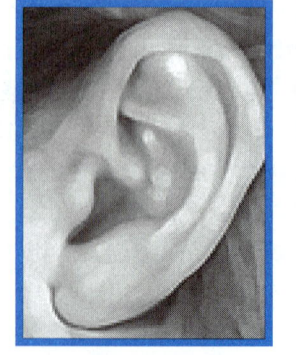

Unattached Lobe   *Attached Lobe*

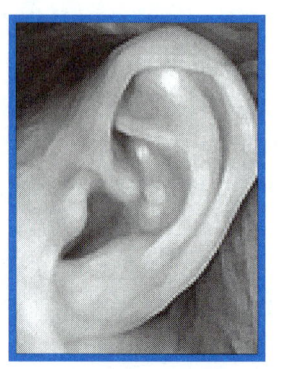

dominant trait   recessive trait

CHAPTER 11: REPRODUCTION AND INHERITED TRAITS 149

APPLYING WHAT YOU HAVE LEARNED

◆ Some pea plants have purple flowers; others have white flowers. Purple flowers are dominant in pea plants. A pea plant inherits a gene for purple flowers from one parent and a gene for white flowers from another parent. What color will its flowers be? _____ Explain your answer. _____

WHAT YOU SHOULD KNOW

☐ You should know that some characteristics, called **traits**, are inherited from an organism's parent or parents. Because organisms reproduce, their **species** continues and **inherited traits** are passed from one generation to the next.

☐ You should know that in **asexual reproduction**, all inherited traits come from a single parent. Asexual reproduction assures genetic continuity and limits the spread of detrimental characteristics.

☐ You should know that in **sexual reproduction**, an egg and sperm unite and the offspring inherits some traits from each parent. The offspring is never identical to both parents. Dominant traits appear over recessive ones. New combinations of traits increase or decrease an organism's chances for survival.

CHAPTER STUDY CARDS

Heredity

★ **Inherited Trait.** A trait an organism inherits from its parents, such as height.

★ **Gene.** The group of DNA material that governs a particular trait.

★ **Dominant Trait.** A trait that appears in an organism if it inherits the gene for that trait from either parent.

★ **Recessive Trait.** A trait that appears in an organism only if it inherits that trait from both its parents.

Reproduction

★ **Asexual Reproduction.** Offspring reproduced from one parent. For example, bacteria copy themselves to reproduce, insuring genetic continuity and limiting unfavorable traits.

★ **Sexual Reproduction.** Two parents, one male and one female, are needed to produce an offspring. Two genes control each trait, and the offspring inherits one of the genes from each parent. This creates a greater variety of genetic combinations.

CHECKING YOUR UNDERSTANDING

1. Pea plants have a covering on their seeds, known as a pod. A green pod is the dominant trait, and a yellow pod is a recessive trait. Which statement must be true for the parents of a plant with a yellow pod?
   A. One parent had two genes for a green pod.
   B. One parent had two genes for a yellow pod.
   C. Both parents had at least one gene for a green pod.
   D. Both parents had at least one gene for a yellow pod.

   LS: B
   G6.6

   HINT: This question tests your understanding of inherited traits. In order to have a yellow pod, a pea plant must inherit a yellow pod gene from each parent. Therefore, each parent must have at least one yellow pod gene. This is the case only in **Choice D**.

   Now try answering some additional questions on your own.

2. For a certain type of cat, the gene for black fur is dominant while the gene for brown fur is recessive. Two cats with black fur produce an offspring with brown fur. Which best describes the genes of the parent cats?
   A. Both parents carry one recessive gene.
   B. Neither parent carries the recessive gene.
   C. One parent carries two recessive genes, but the other does not.
   D. One of the parents carries the recessive gene, but the other parent does not.

   LS: B
   G6.6

3. Where are the instructions for the inherited traits of an organism found?
   A. cell wall
   B. cell membrane
   C. DNA molecules
   D. glucose molecules

   ♦ Examine the Question
   ♦ Recall What You Know
   ♦ Apply What You Know

   LS: B
   G6.4

4. What is an important advantage of asexual reproduction?
   A. It creates new genetic combinations.
   B. It insures there will be genetic continuity.
   C. Each offspring is different from its parents.
   D. It readily adapts to environmental changes.

   LS: B
   G8.1

5. Which is an example of an inherited trait?
   A. Mrs. Smith is five feet tall.
   B. Aiesha has learned about algebra.
   C. Dwayne eats ice cream every night.
   D. Mr. Smith is worried about his daughter's grades.

   LS: B
   G6.7

6. Which is an example of sexual reproduction?
   A. Bacteria cells divide.
   B. A starfish grows a new arm.
   C. An egg is fertilized by a sperm.
   D. A new plant develops from a piece of an old one.

   ♦ Examine the Question
   ♦ Recall What You Know
   ♦ Apply What You Know

   LS: B
   G6.6

Use the following information to answer questions 7 and 9.

A scientist observed bacteria cells under the microscope and carefully recorded her observations.

10:00 am	10:20 am	10:40 am	11:00 am

7. What process was the scientist observing?
   A. photosynthesis
   B. cellular respiration
   C. sexual reproduction
   D. asexual reproduction

   LS: B
   G6.5

8. The scientist compared one of the cells at 11:00 a.m. with a photograph of the cell she saw at 10:00 a.m. Which best describes what she found?
   A. The two cells had different traits.
   B. The two cells shared some traits.
   C. The two cells had identical traits.
   D. The two cells had different learned behaviors.

   LS: B
   G6.5

9. What did the scientist most likely observe at 11:40 A.M.?
   A. 64 cells with similar genes
   B. 8 cells with genes from each parent
   C. 16 cells with different inherited traits
   D. 32 cells with identical inherited traits

   LS: B
   G6.5

10. Which is an example of a characteristic that Carl learned from his parents?
    A. blue eyes
    B. brown hair
    C. long, thin fingers
    D. ability to paint

    LS: B
    G6.7

11. Young spiders spin webs to catch insects to eat. How do young spiders learn how to build their webs?
    A. They practice this skill each day.
    B. They learn from watching other spiders.
    C. They inherit this trait from their parents.
    D. They build webs in response to their environment.

    LS: B
    G6.7

12. When scientists study humans, they see many likenesses between children and their parents. Some of these characteristics are inherited and some are learned.

    In your **Answer Document**, describe one trait that would be inherited by a child from its mother and father. Then describe one likeness that might be learned from its parents. (2 points)

    LS: B
    G6.7

13. No individual organism can live forever. As a result, reproduction is necessary for the continuation of every species.

    In your **Answer Document**, describe two types of reproduction and identify an advantage of both. (4 points)

    LS: B
    G6.4

14. Asexual reproduction and sexual reproduction each have certain advantages for the survival of the species.

    In your **Answer Document**, complete the following chart. (4 points)

    LS: B
    G8.1

	Asexual Reproduction	**Sexual Reproduction**
Describe the process		
Explain one advantage		

CHAPTER 12

ECOSYSTEMS AND ADAPTATION TO ECOLOGICAL CHANGE

In this chapter, you will learn how organisms interact in ecosystems. You will also learn how species adapt, or fail to adapt, to changes in ecosystems.

— MAJOR IDEAS —

A. An **ecosystem** is the community of living organisms in an area. An ecosystem always includes both **biotic** (*living*) resources and **abiotic** (*non-living*) resources, such as light, water and soil.

B. Different species in an ecosystem interact through **symbiotic** relationships. Some species become so adapted to each other that they need one another to survive.

C. Ecosystems recycle both matter and energy. All energy enters the ecosystem through sunlight. Plants convert this solar energy into chemical energy. Some of this chemical energy is transferred to **consumers**, who use it to carry on life functions.

D. Natural occurrences, overpopulation, and human activities can cause changes to ecosystems. Some changes to ecosytems occur slowly, while others may be rapid. A **diversity** of species is developed through gradual processes of change. If a species fails to adapt to change, it may become **extinct**.

WHAT IS AN ECOSYSTEM?

All organisms depend on both the environment and other organisms to survive. **Ecology** is the study of relationships between living organisms and their environment. Ecologists study how all the different organisms in an area, together with the non-living environment, make up an **ecological system**, often known as an **ecosystem**.

TYPES OF ECOSYSTEMS

There are many types of ecosystems. Some are **land ecosystems**. These include forest, grassland, desert, and tundra ecosystems. Ecosystems in water are known as **aquatic ecosystems**. These include freshwater ecosystems like the Great Lakes, and ocean ecosystems like the Continental Shelf, Great Barrier Reef, and the mid-ocean floor.

The Size of an Ecosystem. The size of an ecosystem can vary greatly, from a small pond to a vast forest. Different ecosystems are often separated by geographical barriers, such as deserts, mountains or oceans. However, the borders of ecosystems are not rigid, and one ecosystem often blends into another. An ecosystem, such as a tropical rainforest, may also have several smaller ecosystems within it, such as a forest canopy ecosystem and a forest floor ecosystem.

POPULATIONS IN AN ECOSYSTEM COMMUNITY

All of the organisms found in an ecosystem are called a **community**. For example, all of the trees, bacteria, worms, squirrels, foxes, and deer living in a temperate forest ecosystem make up its community. All of the organisms of the *same species* in a particular ecosystem are known as the **population** of that species. For instance, all of the alligators found in the Everglade wetlands of Florida make up that ecosystem's *alligator population*.

Alligators play an important role in the ecological balance in the Everglades.

APPLYING WHAT YOU HAVE LEARNED

◆ Think of an ecosystem in the area where you live.
  • Identify the ecosystem's location. _____
  • Identify some of the different organisms that live in that ecosystem.
  _____
  _____

ABIOTIC AND BIOTIC RESOURCES

How many organisms an ecosystem can support depends on both its **abiotic** (*non-living*) resources and its **biotic** (*living*) resources.

ABIOTIC RESOURCES

Abiotic (*non-living*) resources often determine what kind of organisms can live in an area. They also affect the ability of organisms to survive and reproduce. These resources include water, humidity, soil, minerals, and sunlight. Differences in the availability of these resources give rise to the different types of ecosystems.

BIOTIC RESOURCES

Biotic (*living*) resources also affect ecosystems. Biotic resources are all the living things that directly or indirectly affect an organism in an ecosystem. All organisms require nutrients and energy to survive. Many species obtain these resources from other living things. Every ecosystem therefore contains different populations of species that interact together.

APPLYING WHAT YOU HAVE LEARNED

◆ For each of the following items, check (✔) if it is a **biotic** or an **abiotic** resource:

Item	Abiotic (non-living)	Biotic (living)
A frog		
A fallen leaf		
A rock		
A piece of wood		

HOW DIFFERENT SPECIES INTERACT

There are many ways that species in an ecosystem interact. Many of these interactions are symbiotic. In a **symbiotic relationship**, one species lives off another. Some species become so adapted to other species in the ecosystem that they could not survive without them.

★ **Predators.** Sometimes, one organism — known as a **predator** — captures and kills another organism, known as its **prey**. Predators like wolves and tigers usually have special characteristics like sharp teeth and the ability to run fast to hunt their prey. Prey also have special physical characteristics, like camouflage or eyes on the sides of their heads, to avoid being caught.

★ **Parasites.** A **parasite** feeds on another organism, known as the **host**. Unlike predators, the parasite does not kill the host, at least not immediately. Some parasites attach themselves to the surface of the host, such as ticks. Others, like tapeworms, live inside the host.

The common flea is a parasite that lives off warm-blooded hosts.

★ **Mutualism.** Sometimes there is a cooperative relationship in which two or more species mutually benefit. For example, flowering plants provide nectar to bees. The bees carry pollen from one plant to the flowers of other plants, leading to cross-fertilization. Both the plants and the bees benefit.

★ **Commensalism.** This refers to a one-sided relationship in which one species benefits without harming the other species. For example, some birds live in holes in trees without affecting the trees.

★ **Competition.** Similar species may compete for the same resources. For example, cows and sheep may eat the same grasses. If one species increases in number, competing species may be reduced.

APPLYING WHAT YOU HAVE LEARNED

◆ Provide one example for each of these four types of interaction:

Interaction	Example
Predator/Prey	
Parasite/Host	
Mutualism	
Commensalism	

THE FLOW OF ENERGY AND MATTER THROUGH AN ECOSYSTEM

Energy and matter — in the form of water, gases, and nutrients — are continually transferred and recycled within an ecosystem. Ecologists are often able to trace the flow of energy and matter through an ecosystem.

PRODUCERS

The basic source of energy for all ecosystems is sunlight. **Producers** capture this energy and turn it into sugars through photosynthesis. In land ecosystems, the producers are green plants. These plants use the energy of sunlight in photosynthesis, but they also require carbon dioxide, water, and nutrients from the soil. In aquatic ecosystems, plants, algae, and microscopic phytoplankton use photosynthesis to obtain energy.

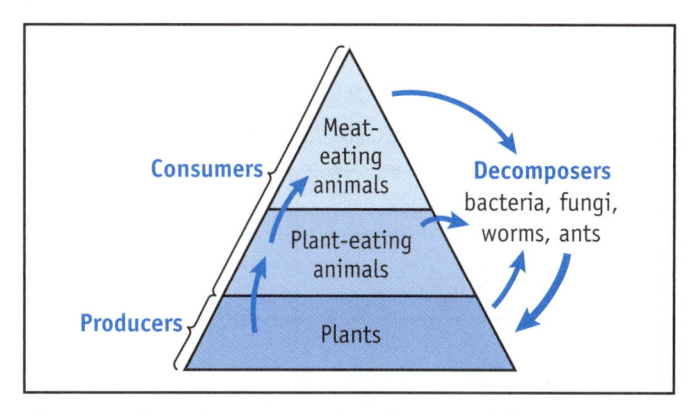

Energy and nutrients are continually recycled in an ecosystem.

CONSUMERS

Unlike plants, animals cannot make their own food. They are **consumers**, which eat other organisms to obtain organic molecules containing energy.

★ Some animals (*herbivores*) eat only plants.

★ Other animals (*carnivores*), such as tigers, only eat other animals.

★ Still other animals (*omnivores*), such as dogs and humans, eat both plants and animals.

DECOMPOSERS

Some organisms, like vultures and ants consume dead organisms or animal wastes. **Decomposers** also include worms, bacteria and fungi. Decomposers are the "garbage disposal" of nature; they take all the dead animals and plants and break them down into nutrients. These nutrients are returned to the soil. Plants then absorb these nutrients and use them to make more food.

This fungus is a decomposer that has been digesting a dead log for several years, eventually turning it into dirt for use by plants.

APPLYING WHAT YOU HAVE LEARNED

♦ Classify each organism as either a **producer**, **consumer** or **decomposer**, and explain your classification.

Organism	Classification	Explanation
Oak tree		
Beetle		
Black bear		

FOOD CHAINS AND WEBS

A **food chain** is a diagram tracing the flow of energy and nutrients through a single ecosystem. It shows the specific links between organisms in the ecosystem. For example, the grass (*producers*), rabbits (*consumers*), and coyotes (*consumers*) on a prairie form a single food chain.

Grass → Rabbits → Coyotes

Energy flows from the sun to the grasses. Rabbits feed on the grasses and absorb their energy and nutrients. Coyotes eat some of the rabbits and absorb their energy and nutrients.

A **food web**, like the one on the next page, shows several related food chains. When the rabbits and coyotes die, worms, bacteria, and fungi decompose their bodies and enrich the soil with nutrients for the grasses. Water, carbon dioxide, and oxygen are similarly recycled through the ecosystem.

For example, water is stored in the ocean, in freshwater sources, and underground. Some of this water evaporates into the atmosphere. Later, this water comes back to Earth's surface as rain, snow, sleet or fog. Plants take this water through their roots, and animals drink the water. All these organisms then release water back into the environment. Plants release water through pores in their leaves, and animals perspire. The water is continually recycled through biotic and abiotic elements of the ecosystem.

THE FOOD WEB IN A PRAIRIE ECOSYSTEM

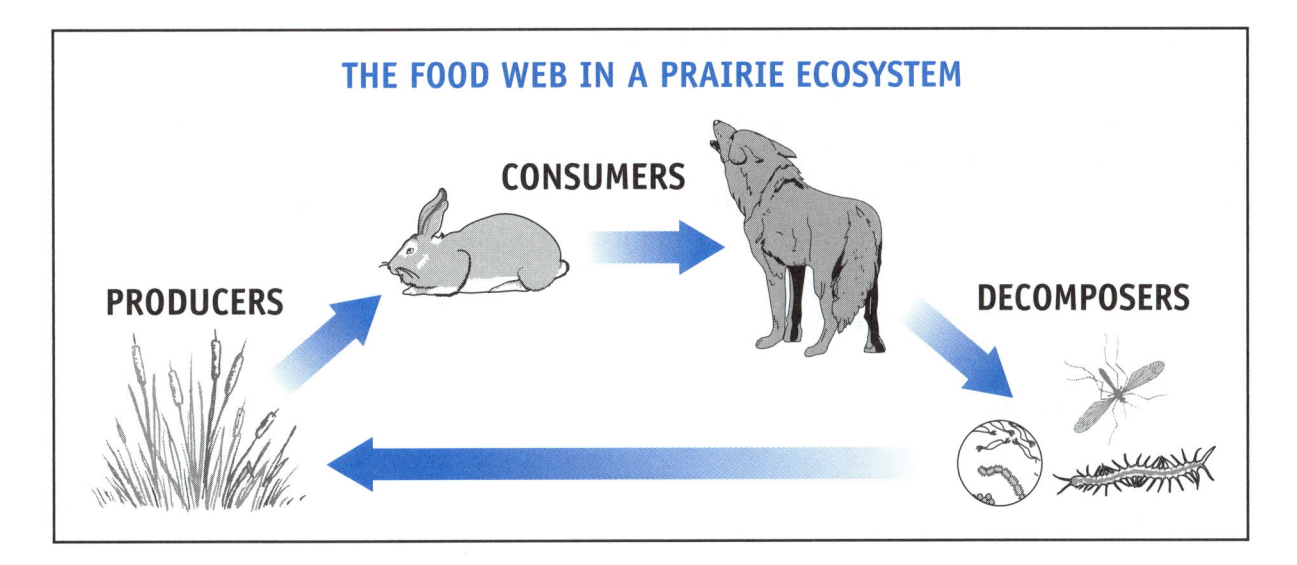

PRODUCERS → CONSUMERS → DECOMPOSERS

APPLYING WHAT YOU HAVE LEARNED

◆ On a separate sheet of paper, draw a food chain showing yourself as one of the consumers. In your diagram, show what you ate for dinner last night and consider where the energy in those ingredients came from.

CHANGES IN ECOSYSTEMS

The existence of limited resources (*water, food, oxygen, sunlight, and space*), competition with other species, and the loss of individuals to predators limits the growth of each species in an ecosystem. There may be periodic changes within the community of an ecosystem, but a delicate **equilibrium** (*balance*) is usually reached among its different species.

POND ECOSYSTEM

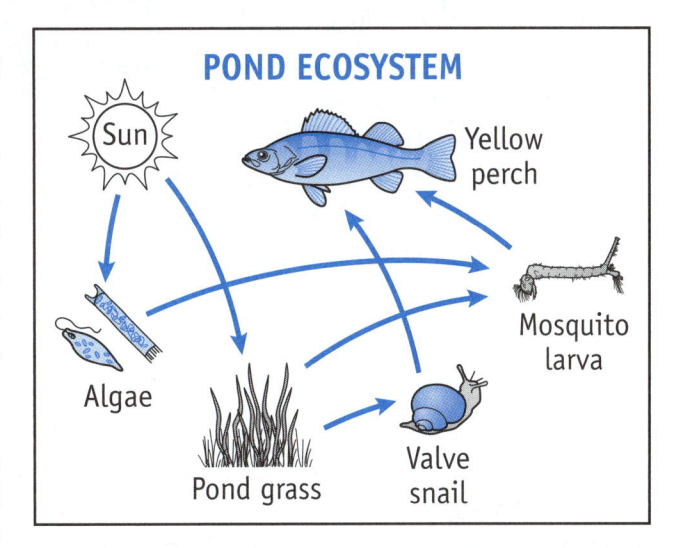

Sun, Algae, Pond grass, Valve snail, Mosquito larva, Yellow perch

For example, in a pond ecosystem, the sun provides energy to algae and pond grass. Snails and mosquito larvae eat the grass and algae. Fish eat the larvae. When the grass, snails, fish, and larvae die, they sink to the bottom of the pond where they decompose. Left undisturbed, the pond can continue recycling its resources for hundreds or even thousands of years.

Equilibrium in an Ecosystem. The equilibrium of an ecosystem can be upset by the increased growth of one of its species, by environmental change, by human activity, or by natural disasters. Such changes can have effects that ripple through the ecosystem. For example, wolves may keep the number of deer in a forest ecosystem in check. If human hunters kill the wolves, the deer population may expand too rapidly, threatening plant life and other plant-eating animals in the ecosystem.

APPLYING WHAT YOU HAVE LEARNED

◆ What are some of the forces that help to keep a pond ecosystem in balance? _____

◆ What events might cause a pond ecosystem to change? _____

OVERPOPULATION

If the deer population expands too quickly, the ecosystem will become overpopulated. The increased number of deer in the ecosystem will eat more shrubs and grass. This will leave less food available for the other animals. Competitors for the same resources, like rabbits, may become affected. As the deer population continues to multiply, they will find there are insufficient shrubs and grasses available even for themselves. Some deer may migrate to new areas. Other deer will starve to death when they can no longer find food. Eventually, the number of deer in the ecosystem will decrease again.

When deer overpopulate it leaves less food for other animals.

ENVIRONMENTAL CHANGE

Over time, the environment of an ecosystem may gradually change. For example, a region may become cooler or drier. Species of organisms must adapt to these changes in order to survive. The process, known as **natural selection**, helps explain how species adapt to environmental change. Individuals of the same species with favorable inherited traits will generally survive the changes and reproduce. Those with less favorable traits generally will not.

For example, some rabbits have naturally thick fur, while others have thin fur. This is a difference that may have occurred as a result of accidental changes to genes. If the climate of a region changes and becomes colder, the rabbits with thick fur will have a better chance of surviving than the rabbits with thin fur. The rabbits with thicker fur will better withstand cold temperatures and have more time to eat. More of the thick-furred rabbits will survive and reproduce than of the thin-furred rabbits. Gradually, they will make up a larger proportion of the rabbit population in that ecosystem.

Adapting to Change. Random genetic differences thus become important in times of environmental change. Organisms with favorable traits are more likely to survive and reproduce, passing their favorable genetic traits on to their offspring. Other organisms from the same species, with unfavorable traits for the new environment, will tend to die off without children. Species thus adapt to the new conditions very gradually, over many generations. Eventually, all of the surviving offspring in the species will have those favorable traits. Scientists refer to the fact that a species adapts to its environment through random genetic change and natural selection as the **theory of evolution**. Most scientists believe that this process helps to explain the great diversity of life now found on our planet.

Dandelions adapted to their environment by producing white fluff on their seeds to increase their survival.

SEXUAL REPRODUCTION AND ADAPTATION

In the last chapter, you learned that sexual reproduction continually introduces new combinations of inherited traits. As a result, there is usually a greater variety of characteristics among organisms that have sexual reproduction than among those that reproduce asexually. Some of the new combinations that occur may not be suited to their environment. The new traits may decrease the offsprings' chances for survival. However, other new combinations of traits may be very well suited to the environment. These organisms will have a better chance of surviving and reproducing. Because of the greater variety among organisms with sexual reproduction, this method of reproduction is often very successful in adapting to changes in the environment.

APPLYING WHAT YOU HAVE LEARNED

In the 1830s, a young naturalist, **Charles Darwin**, traveled to a group of islands off the coast of South America. Darwin found a wide variety of finches, a type of bird, with different beaks. Darwin found that some had narrow beaks that could fit into the gaps between stones to find insects. Others had large beaks useful for cracking shells and nuts found on the island. Each type of beak appeared well-suited to finding and using a different food source.

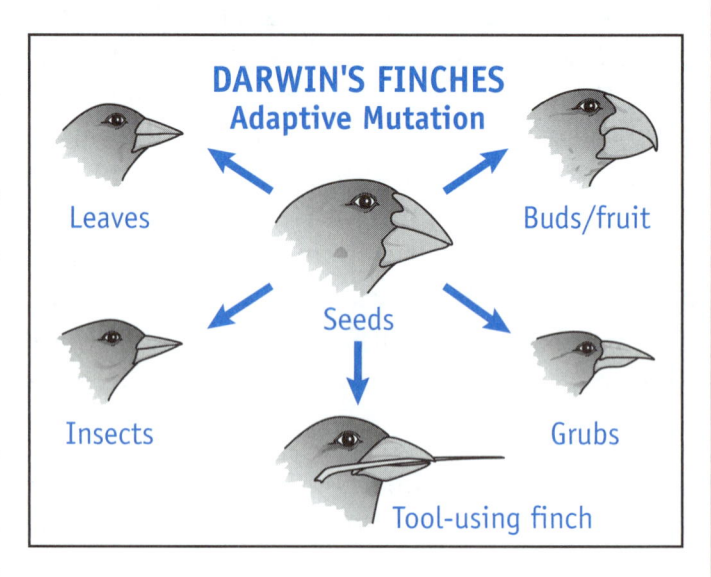

DARWIN'S FINCHES
Adaptive Mutation
Leaves — Buds/fruit — Seeds — Insects — Grubs — Tool-using finch

◆ What would help explain these differences if all these birds came from a common ancestor? _____

◆ How does sexual reproduction help a species adapt more easily to environmental change? _____

HUMAN ACTIVITY

Human activity can have important consequences on an ecosystem. For example, when humans cut down trees to build homes, this may deprive some animals of their natural habitat. Forced to find a new environment to live in, these animals may fail to adapt and die out.

Human Impact on Ecosystems. Other changes caused by humans are affecting ecosystems across the globe. Air and water pollution are poisoning some species and raising average temperatures. Oil spills affect ocean and coastal life. Logging, deforestation, and mining are destroying many natural habitats.

An oil spill in Alaska had a large impact on coastal ecosystems.

NATURAL DISASTERS AND ECOLOGICAL SUCCESSION

Natural disasters as much as human activity can bring rapid changes to an ecosystem. Such natural disasters include fires, storms, and droughts. These drastic environmental events often lead to a **succession** (*series*) of changes in an ecosystem. Different organisms appear at each stage as the environment changes. For example, a fire may destroy the ancient trees of a temperate forest ecosystem. Weeds and grasses will quickly spring up in the ashes, which contain valuable nutrients. Next, shrubs and plants will grow, blocking the light for the grasses. Finally, pine trees and trees that shed their leaves in fall will take root, replacing the shrubs. Many shrubs will die because the trees block their sunlight. A new temperate forest develops, beginning the cycle all over again.

After fire destroys this forest, a succession of changes will occur.

APPLYING WHAT YOU HAVE LEARNED

◆ Identify one ecosystem near where you live, and describe some of the effects human activity has had on that ecosystem. _____
_____

◆ What steps do you think need to be taken to preserve this ecosystem?
_____

RAPID CHANGES AND EXTINCTION

You have learned that species often adapt to environmental change through genetic changes and natural selection. This evolutionary process generally takes millions of years. Sometimes, however, natural selection may fail to work. The environment may change so quickly that a species is unable to adapt. For example, a large fire may destroy a forest, or "global warming" may change a grassland into a desert. In such cases of rapid change, an entire species may disappear from Earth.

Many scientists believe that about 65 million years ago, a giant meteor crashed into Earth. Dust and smoke from the impact covered Earth's atmosphere, blocking much of the sunlight. Many plants died in the darkness. After living on Earth for about 165 million years, the dinosaurs that has fed on these plants entirely died out. When a species fails to adapt to change and all of its members die out, scientists say the species has become **extinct**. A species that is close to extinction is known as an **endangered species**. Today, many species are endangered by the effects of human activities.

Scientists believe the dinosaurs may have become extinct because they failed to adapt to a natural disaster.

Fossil Remains. Sometimes, a dead plant or animal leaves behind bones, shells, leaves or tracks. For example, a dinosaur might step in mud, leaving its footprint. The mud dries, and sand settles on the mud footprint. The sand and mud harden into different types of rock. The footprints can still be seen when the rock is dug up. By examining such **fossils**, scientists can tell what many plants and animals looked like millions of years ago. For example, scientists can see the leaves of ancient ferns, and the skeletons and shapes left by dinosaurs. From these fossils, scientists can not only learn about dinosaurs, but they can identify approximately when they died out.

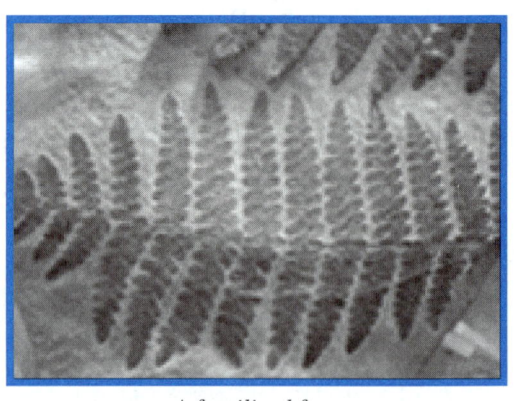

A fossilized fern.

APPLYING WHAT YOU HAVE LEARNED

Some scientists believe that more species are becoming extinct today than at any other time since the end of the dinosaurs 65 million years ago.

◆ What factors might explain this? _____
_____

◆ If an asteroid were to hit Earth tomorrow, might humans become extinct?
_____

CHAPTER 12: ECOSYSTEMS AND ADAPTATION TO ECOLOGICAL CHANGE

WHAT YOU SHOULD KNOW

☐ You should know that an **ecosystem** is the community of living organisms in an area. An ecosystem always includes both **biotic** (*living*) resources and **abiotic** (*non-living*) resources, such as light, water and soil.

☐ You should know that different populations in an ecosystem interact through **symbiotic** relationships. These include **predation**, **parasitism**, **mutualism**, and **commensalism**. Some species become totally adapted to each other.

☐ You should know that ecosystems transfer and recycle both matter and energy. All energy enters each ecosystem through sunlight. Plants convert this solar energy into chemical energy. Some of this chemical energy is transferred to **consumers**, who use it to carry on life functions.

☐ You should know that **natural occurrences**, **overpopulation**, and **human activities** can cause changes to ecosystems. Some changes to ecosystems occur slowly, while others may be rapid. A **diversity** of species develops through gradual processes of adapting to change. If a species fails to adapt to change that is too rapid, the species may become **extinct**.

CHAPTER STUDY CARDS

Ecosystems

An **ecosystem** is the community of all living organisms and their environment in a specific area.

★ **Nonliving (abiotic) Environmental Factors.** Non-living factors influencing an ecosystem include water, sunlight, and soil.

★ **Community.** All the organisms found in a single ecosystem.

★ **Population.** All of the organisms of the same species in a particular ecosystem.

★ **Examples of Land Ecosystems:** Temperate forest, tropical rain forest, grassland, desert, and tundra.

Flow of Resources in an Ecosystem

★ **Interaction of Organisms.** Predators, parasites, mutualism, commensalism, and competition.

★ **Recycling of Energy and Nutrients.**
  • **Producers** (*plants*) obtain energy from sunlight, water and nutrients from soil.
  • **Consumers** (*animals*) eat plants or other animals; provide nutrients and CO_2 to be used by plants.
  • **Decomposers** (*bacteria, fungi*) break down dead organisms into organic compounds.

★ **Ecological Succession:** drastic events, like fire, bring a series of changes to an ecosystem.

CHECKING YOUR UNDERSTANDING

1. Which of the illustrations below represents an ecosystem?

   LS: C
   G7.3

   A   B   C   D

   HINT: This question examines your understanding of an ecosystem. Recall that an ecosystem consists of all of the living organisms in an area, together with their nonliving environment. Although **Illustration A** shows different species, it does not represent an ecosystem of organisms and their nonliving environment. Only **Illustration D** has both the organisms and their nonliving environment, both of which are necessary for an ecosystem.

   Now try answering some additional questions on your own.

2. Which can be described as a species population in an ecosystem?
   A. all the honey bees in an orchard
   B. all the plants and animals in a pond
   C. all the nonliving conditions in an area
   D. all living things in the Earth's atmosphere

   LS: C
   G7.2

3. Which evidence best demonstrates that different organisms in an ecosystem depend on each other?
   A. Parasites often kill their hosts.
   B. Plants are able to convert light into chemical energy.
   C. Different environments support different species of organisms.
   D. Animals eat plants while plants need nutrients from decomposed animals.

   LS: C
   G7.2

Use the diagram below to answer question 4.

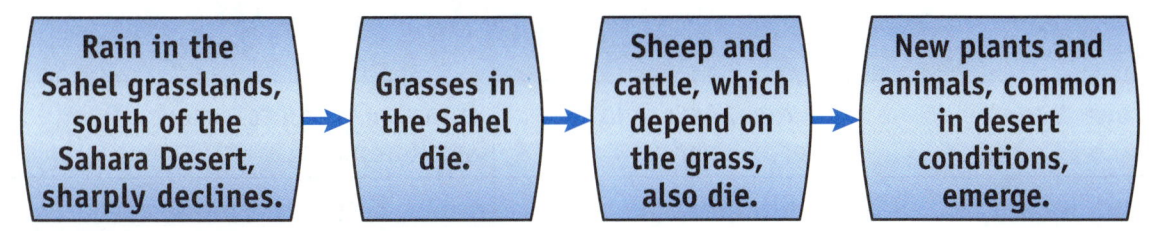

| Rain in the Sahel grasslands, south of the Sahara Desert, sharply declines. | → | Grasses in the Sahel die. | → | Sheep and cattle, which depend on the grass, also die. | → | New plants and animals, common in desert conditions, emerge. |

4. What does this diagram illustrate?

   A. the extinction of a species
   B. equilibrium in an ecosystem
   C. a succession of ecological changes
   D. the recycling of nutrients in an ecosystem

LS: D
G7.5

Use the passage below to answer questions 5 and 6:

SNOWSHOE HARES

The snowshoe hare is the prey of many woodland animals. Foxes, coyotes, and weasels hunt the hares by day, while owls hunt them at night. The snowshoe hare survives partly through changing fur color, creating a camouflage. Some favorite foods of the hare during the winter months are twigs of maple, birch, and apple trees. Grasses and clover replace this diet during the spring and summer. Hares tend to feed during the hours of dusk and dawn, when the light is low and their predators are inactive.

The Snowshoe Hare

5. Based on its diet, how should the snowshoe hare be classified?

   A. predator
   B. producer
   C. consumer
   D. decomposer

   ♦ Examine the Question
   ♦ Recall What You Know
   ♦ Apply What You Know

LS: C
G7.7

6. Since foxes, coyotes, and weasels hunt snowshoe hares, how should these hunting animals be classified?

   A. predators
   B. parasites
   C. producers
   D. decomposers

LS: C
G7.2

Use the information in the table below to answer question 7.

Species of Bear	Location	Food Sources
Brown Bear	Europe, Asia, and North America	fruits, nuts, roots, insects, fish, small animals
Black Bear	North America	fruits, nuts, plant roots, bee honey, insects, rats, mice, fish
Polar Bear	Arctic	seals, fish, sea birds, hares, caribou, musk oxen
Panda Bear	China	bamboo stems and leaves

7. Which of these four species of bear would be most in danger of becoming extinct if environmental changes made some of their food sources unavailable?

   A. polar bear
   B. black bear
   C. brown bear
   D. panda bear

   LS: D
   G8.5

8. Which of the following best completes the food web illustrated on the right?

   A. trees
   B. fish
   C. rabbit
   D. hawk

   LS: C
   G7.2

   ◆ Examine the Question
   ◆ Recall What You Know
   ◆ Apply What You Know

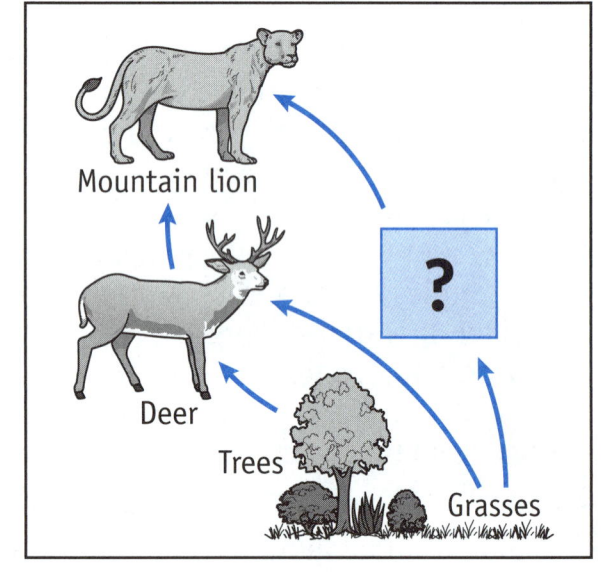

9. What is the basic source of energy for all ecosystems?

   A. solar energy, which is converted by plants
   B. nuclear energy, which is stored in Earth's resources
   C. electrical energy, which comes from water vapor in the air
   D. kinetic energy, which is produced by the motion of animals

   LS: C
   G7.7

10. In an ecosystem, some species become so adapted to each other that neither could survive without the other.

    In your **Answer Document**, select one species and describe how it has become so adapted to organisms of another species that it could not survive without them. (2 points)

    LS: C
    G7.2

CHAPTER 12: ECOSYSTEMS AND ADAPTATION TO ECOLOGICAL CHANGE

CONCEPT MAP OF LIFE SCIENCES

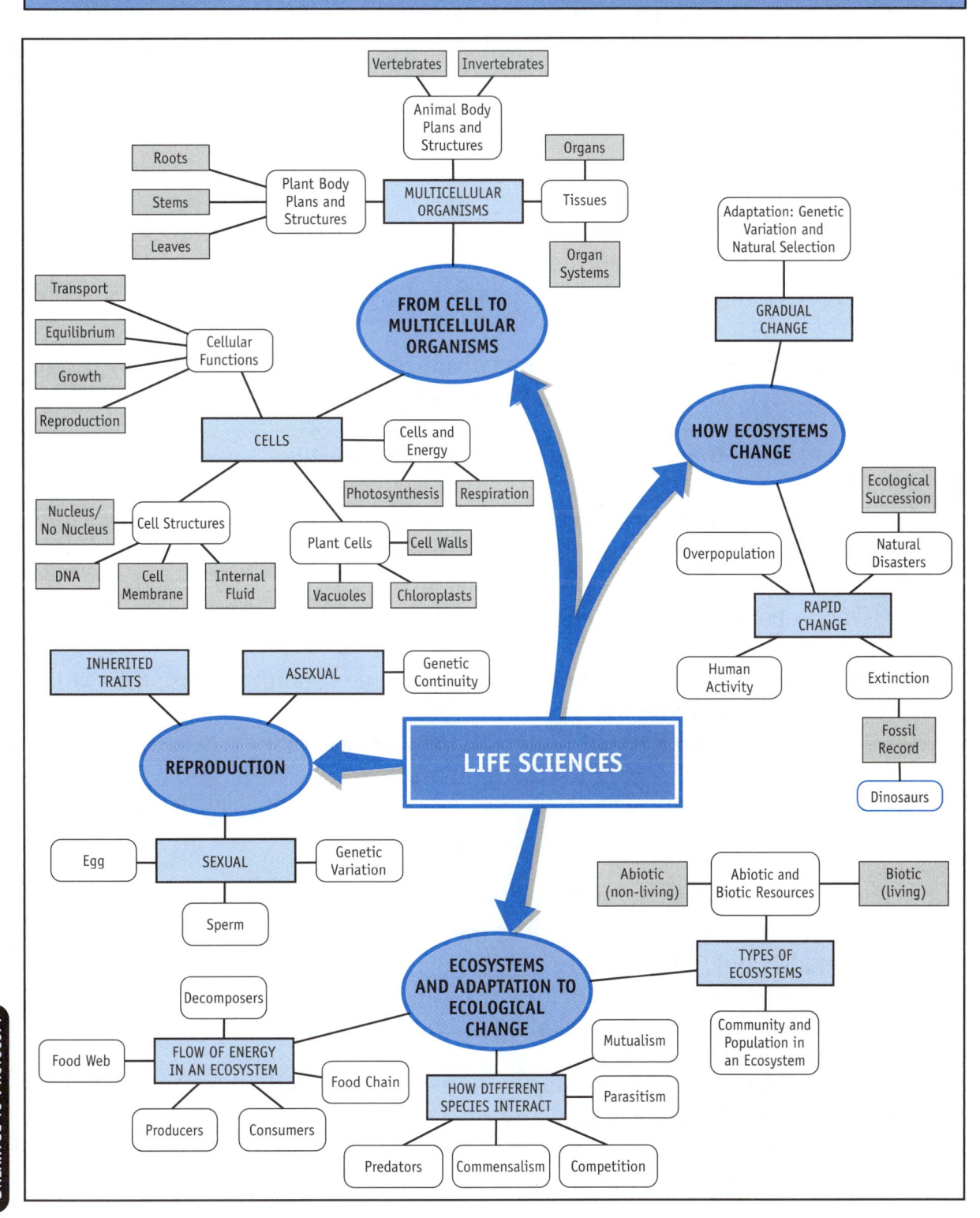

LIFE SCIENCES

FROM CELL TO MULTICELLULAR ORGANISMS

- **MULTICELLULAR ORGANISMS**
  - Animal Body Plans and Structures
    - Vertebrates
    - Invertebrates
  - Plant Body Plans and Structures
    - Roots
    - Stems
    - Leaves
  - Tissues
    - Organs
    - Organ Systems

- **CELLS**
  - Cellular Functions
    - Transport
    - Equilibrium
    - Growth
    - Reproduction
  - Cells and Energy
    - Photosynthesis
    - Respiration
  - Cell Structures
    - Nucleus/No Nucleus
    - DNA
    - Cell Membrane
    - Internal Fluid
  - Plant Cells
    - Cell Walls
    - Vacuoles
    - Chloroplasts

HOW ECOSYSTEMS CHANGE

- **GRADUAL CHANGE**
  - Adaptation: Genetic Variation and Natural Selection
- **RAPID CHANGE**
  - Overpopulation
  - Ecological Succession
  - Natural Disasters
  - Human Activity
  - Extinction
    - Fossil Record
      - Dinosaurs

REPRODUCTION

- **INHERITED TRAITS**
- **ASEXUAL**
  - Genetic Continuity
- **SEXUAL**
  - Egg
  - Sperm
  - Genetic Variation

ECOSYSTEMS AND ADAPTATION TO ECOLOGICAL CHANGE

- **TYPES OF ECOSYSTEMS**
  - Abiotic and Biotic Resources
    - Abiotic (non-living)
    - Biotic (living)
  - Community and Population in an Ecosystem
- **FLOW OF ENERGY IN AN ECOSYSTEM**
  - Decomposers
  - Food Web
  - Food Chain
  - Producers
  - Consumers
- **HOW DIFFERENT SPECIES INTERACT**
  - Mutualism
  - Parasitism
  - Predators
  - Commensalism
  - Competition

169

TESTING YOUR UNDERSTANDING

1. A new plant appears with a different flower color than nearby plants of the same species. What would be the most likely cause of its appearance?
   A. photosynthesis
   B. sexual reproduction
   C. asexual reproduction
   D. gradual environmental change

2. A large population of houseflies was sprayed with an insecticide. However, a great number of houseflies were unaffected by this insecticide. What would best explain this situation?
   A. A species' traits tend to remain unchanged.
   B. The houseflies were able to avoid the insecticide.
   C. The environment for houseflies changed favorably.
   D. Variations in traits permitted many of the houseflies to survive.

3. Some rabbits have white fur while other rabbits in the same area have brown fur. If the climate of the area changed so that there was increased snow in winter, which would be most likely to happen?
   A. The proportion of white and brown rabbits would stay about the same.
   B. More brown rabbits would survive than white rabbits because they would find each other more quickly.
   C. More white rabbits would survive than brown rabbits because they are better able to withstand the cold temperatures.
   D. More white rabbits would survive than brown rabbits because predators would not see them as easily against the snow.

4. What would happen to animals if all the plants in the world died out?
   A. Animals would survive by eating minerals.
   B. Animals would obtain energy by breaking down water.
   C. Animals would also eventually die out from lack of food.
   D. Animals would learn to make their own food from sunlight.

5. Human beings eat chickens, turkeys, fish, cows, and pigs. What type of relationship does this illustrate?
   A. mutualism
   B. predator – prey
   C. parasite – host
   D. commensalism

6. In a grasslands ecosystem, lions are predators that eat antelope. What impact would the sudden disappearance of lions have on this ecosystem?
   A. The grasses would grow larger.
   B. More antelopes would eat the grasses.
   C. Antelopes would start to eat each other.
   D. The climate would grow increasingly warmer.

LS: C
G6.8

Use the information in the box below to answer question 7.

The dodo bird once lived on a large island off the coast of Africa. It lived on this island with plenty of food and no predators. It lived and nested on the ground and survived on fruits that had fallen from trees. Dodos were large, but their wings and breast muscles were too small for them to fly. When humans arrived on the island, they hunted dodo birds. Dogs and pigs, which humans brought to the island, attacked dodo nests and ate their eggs. The last dodo bird lived around 1690.

The dodo bird.

7. What conclusion can be drawn from this evidence?
   A. Dodo birds failed to adapt to rapid climatic change.
   B. Human activity has led to the extinction of some species.
   C. The island's ecosystem was overpopulated with dodo birds.
   D. Like dinosaurs, dodo birds became extinct after a natural disaster.

LS: D
G8.5

8. Which is an example of mutualism?
   A. Ticks live off the blood of deer.
   B. Humans only hunt deer at special times of the year.
   C. Bees take pollen from flowers and spread it to other flowers.
   D. Vines are long-stemmed plants that wrap around other plants.

LS: C
G7.2

9. What structure is found in a plant cell but not in an animal cell?
   A. a nucleus   C. DNA molecules
   B. chloroplasts   D. a cell membrane

LS: A
G6.3

10. Corals get about 90 percent of their food from algae that live inside coral tissues. The algae provide carbohydrates for the corals. The corals provide shelter and nitrogen for the algae. What type of relationship does this illustrate?
    A. parasitism   C. mutualism
    B. predation   D. commensalism

LS: C
G7.2

11. A scientist examines the shell of an American lobster. The scientist observes that the lobster has two large claws, two small claws, two antennae, a tail, and six walking legs. What do these observations show?

   A. The lobster is a vertebrate.
   B. The lobster has a radial body plan.
   C. The lobster has no internal structures.
   D. The lobster has a symmetrical body plan.

12. Which correctly shows the order of structures from simplest to most complex?

   A. cell → tissue → organ system → organ
   B. tissue → cell → organ → organ system
   C. organ system → tissue → cell → organ
   D. cell → tissue → organ → organ system

13. How does the cell of a multicellular organism differ from the cell of a unicellular organism?

   A. It cannot move.
   B. It obtains all of its energy from sunlight.
   C. It must meet all of the needs of the organism.
   D. It carries on more highly specialized functions.

14. How do animal cells dispose of their wastes?

   A. by sending them into vacuoles
   B. by copying their DNA molecules
   C. by expelling them outside the cell
   D. by converting them into chemical energy

15. Which process is almost the reverse of photosynthesis?

   A. parasitism
   B. extinction
   C. cellular respiration
   D. asexual reproduction

16. The large number of needles on a Douglas fir tree can prevent most of the sunlight from reaching the floor of a forest. Which organisms would be most affected by this situation?

   A. worms and ants
   B. squirrels and owls
   C. grass and shrubs
   D. mountain lions and foxes

17. Crayfish are found throughout North America in many freshwater communities. Crayfish find shelter under rocks, in plants, or in the sediment at the bottom of lakes. Crayfish eat insects and snails. Sometimes crayfish will also eat plants and algae. Which best describes the role of the crayfish?

   A. parasite
   B. producer
   C. consumer
   D. decomposer

CHAPTER 12: ECOSYSTEMS AND ADAPTATION TO ECOLOGICAL CHANGE 173

Use the information in the diagram to answer question 18.

18. The diagram shows the bones in the limbs of three different organisms. What hypothesis is best supported by differences in the bone arrangements of these organisms?
    A. They are members of the same species.
    B. They all contain the same DNA molecules.
    C. They all descended from the same ancestor.
    D. They have adapted to different environments.

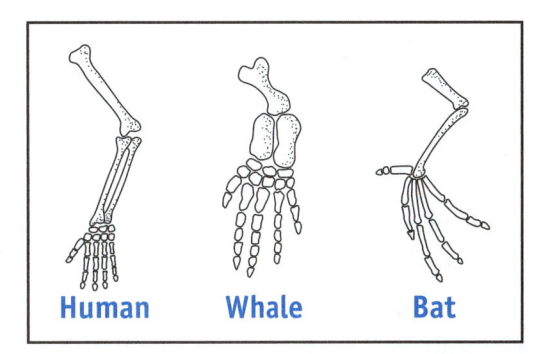

Human   Whale   Bat

LS: D
G8.4

---

19. Natural occurrences or human activities may dramatically affect conditions in Earth's ecosystems.

    In your **Answer Document**, identify one natural occurrence or one human activity that has affected one of Earth's ecosystems.

    LS: C
    G7.6

    Then show how this occurrence or activity led to either the adaptation or extinction of some species. (2 points)

CHECKLIST OF LIFE SCIENCE BENCHMARKS

☐ You should be able to explain that the basic functions of organisms are carried out in cells, that groups of specialized cells form tissues and organs; and that combination of specialized cells make up multicellular organisms that have a variety of body plans and internal structures. [Chapter 10]

☐ You should be able to describe the characteristics of an organism in terms of a combination of inherited traits and recognize reproduction as a characteristic of living organisms essential to the continuation of each species. [Chapters 11 and 12]

☐ You should be able to explain how energy entering an ecosystem as sunlight supports the life of its organisms through photosynthesis, and how energy is transformed through the interactions of organisms and the environment. [Chapter 12]

☐ You should be able to explain how the extinction of a species occurs when the environment changes and its adaptive characteristics are insufficient to allow survival (as seen in evidence of the fossil record). [Chapter 12]

UNIT 5: EARTH AND SPACE SCIENCES

For countless centuries, people have gazed at the night sky and wondered what they were seeing. Today, scientists continue to study the stars and planets, including our own, to understand our place in the universe.

In this unit, you will learn about outer space and our planet, Earth. You will study the processes that occur on our planet and the relationship of our planet to the rest of the universe.

The planet Earth as seen from the moon.

CHAPTER 13: THE UNIVERSE
In this chapter, you will learn about astronomy — the study of stars, the planets, and space. You will learn about galaxies, stars, comets, and asteroids, as well as about the motions of Earth and the moon in space.

CHAPTER 14: EARTH'S INTERIOR AND LANDFORMS
In this chapter, you will learn about the interior of Earth, about tectonic plate motion, and about the constructive and destructive forces that shape Earth's surface. You will also learn about rocks and how they are formed.

CHAPTER 15: THE INTERACTION OF EARTH'S SYSTEMS
In this chapter, you will learn about Earth's lithosphere, hydrosphere and atmosphere. You will see how matter and energy in these three "spheres" interact to recycle water and other materials, and to produce Earth's weather.

CHAPTER 13

THE UNIVERSE

Astronomy — the study of the stars, planets, and outer space — is one of the oldest branches of science. In this chapter, you will learn about stars, galaxies, and our own solar system.

— MAJOR IDEAS —

A. The **universe** consists of billions of galaxies, cosmic gases and dust, and empty space. A **galaxy** is a group of stars and other particles held together by gravity. Scientists classify galaxies by their shapes.

B. **Stars** are balls of superheated gases that produce energy through nuclear reactions. Each star passes through its own life cycle as its temperature and processes change.

C. Scientists measure distances between stars in **light-years**.

D. Our **solar system** consists of the sun, its planets and their moons, dwarf planets, **asteroids**, **comets** and other particles.

E. Objects in the solar system move in regular and predictable motions. Their movements explain such events as years, day and night, the seasons, eclipses, tides and the phases of the moon. The force of **gravity** largely determines the motions of these bodies in space

F. Scientists use a variety of tools to study space, including telescopes, probes, satellites, and manned spacecraft.

THE UNIVERSE

The **universe** is everything that exists — all matter, energy, and space. The universe extends in every direction as far as we can detect. It may go on forever, or it may in fact be limited in size. Scientists have no way of knowing for certain. Most scientists believe that the present universe is expanding. Many scientists also believe the universe began in one place at a single point in time almost 15 billion years ago.

Today, our universe consists of galaxies of stars, clouds of cosmic gas and dust, and enormous distances of empty space. Because of the gigantic distances between stars and galaxies, scientists measure these large distances in **light-years** — the distance that light can travel in one year.

HOW LONG IS A LIGHT-YEAR?

Four hundred years ago, an Italian scientist, Galileo Galilei, believed that light travels from one point to another. He designed an experiment in which an assistant flashed a lantern from a nearby mountain at a certain time. Galileo hoped to record the time it took the light to reach him. In fact, light travels so fast that, for purposes of this experiment, the light was instantaneous. Light does, however, travel. For example, it takes light from the sun 8.31 minutes to travel 93 million miles to Earth. To give you some idea of this distance, if you drove nonstop at 60 mph, it would take you 180 years to go that far. Sunlight does this in about 8 minutes.

Light from the moon takes about 1.2 seconds to reach Earth. So you see the moon as it looked just over a second ago.

How far does light actually travel in one year? It travels 5,879,000,000,000 miles. In the metric system, this is 9,460,730,472,580 km. That is about 63,000 times the distance between Earth and the sun. A **light-year** is a unit representing the distance light can travel in one year. Light-years are used to measure the

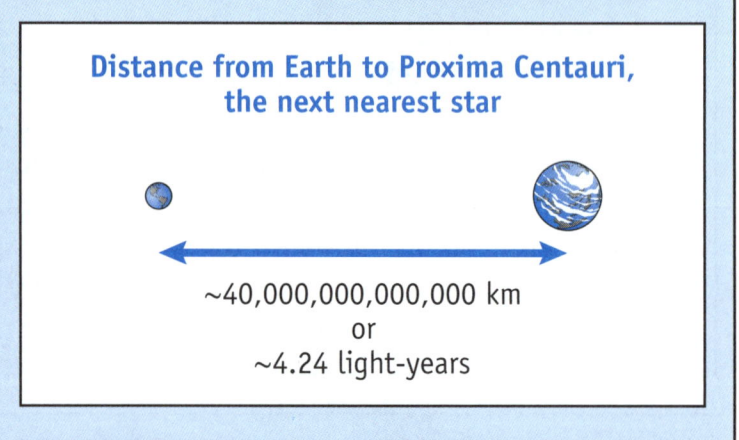

Distance from Earth to Proxima Centauri, the next nearest star

~40,000,000,000,000 km
or
~4.24 light-years

distance between stars because these distances are so vast. For example, the nearest star to Earth after the sun, Proxima Centauri, is 4.24 light-years away. Our galaxy, the Milky Way, is about 100,000 light-years across. The next nearest galaxy — the Andromeda Galaxy — is about 2.5 million light-years away.

STARS

Stars are enormous balls of superheated gases. The sun is the closest star to Earth. Scientists use powerful telescopes to study stars. By analyzing the light rays emitted by a star, scientists are able to tell its composition.

CHAPTER 13: THE UNIVERSE

Most stars are mainly made of two elements: **hydrogen** and **helium**. Each star is actually like a giant nuclear reactor, producing energy through **nuclear fusion** — the joining together of atomic nuclei to make new atoms.

FORMATION

Scientists believe that stars first form out of clouds of gases and dust in space known as **nebula**. The force of gravity pulls the gas and dust together. As the cloud becomes more concentrated, it also begins to spin. Small particles come closer and closer together, until they form a star. The center of a star is extremely hot and dense. Tremendous pressure causes single protons (*hydrogen nuclei*) to fuse together into helium. This process unleashes tremendous amounts of energy. Energy moves from the center of the star outwards. The energy finally radiates from the surface of the star across open space at the speed of light.

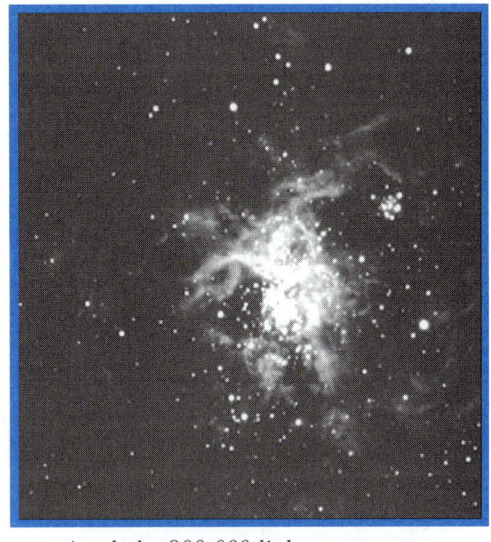

A nebula, 900,000 light years across and 500,000 times the mass of the sun.

LIFE CYCLE OF A STAR

Eventually, a star will turn all its hydrogen into helium. Without the pressure of nuclear fusion pushing the star outward, gravity causes the star to contract. Next, the internal temperature of the star rises, causing it to expand again and become a **red giant**. Gradually, the star begins to cool down. When fusion stops and a large star collapses into itself, it experiences a violent explosion. Heavier elements like iron are formed at this time. Eventually, the star cools down again and becomes a **dwarf** or a **black hole** — an area of dense matter that attracts other matter by its strong gravitational force.

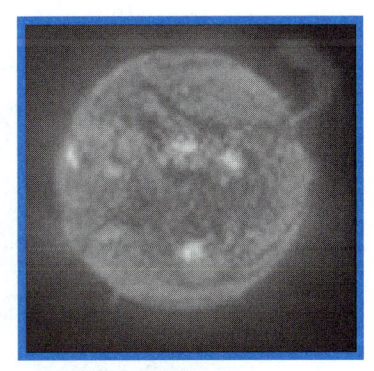

The sun.

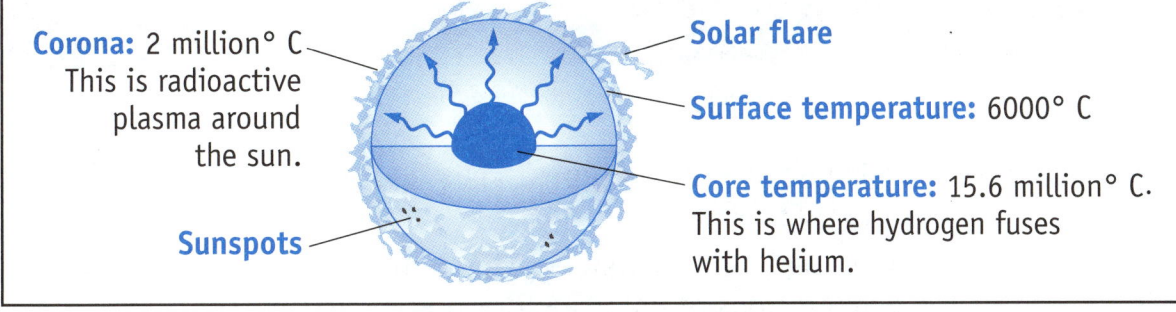

Corona: 2 million° C. This is radioactive plasma around the sun.

Sunspots

Solar flare

Surface temperature: 6000° C

Core temperature: 15.6 million° C. This is where hydrogen fuses with helium.

The sun is 70% hydrogen and 28% helium.

> ## APPLYING WHAT YOU HAVE LEARNED
>
> ◆ Describe how stars are formed and how they produce energy. _____
> _____
> _____
>
> ◆ On a separate sheet, make a timeline showing the life cycle of a typical star.

GALAXIES

Stars are grouped together into **galaxies**. A galaxy, like our own **Milky Way**, will often have billions of stars as well as nebulae and black holes. Each galaxy is thousands of light-years in width. Scientists believe that there are more than 100 billion galaxies in the universe. Galaxies come in many different shapes. The three main types of galaxies are *elliptical*, *spiral*, and *irregular*.

★ An **elliptical galaxy** has older stars. It looks like either a sphere or oval.

★ A **spiral galaxy** has a central sphere of older stars, surrounded by a spinning disk or spiraling arms of nebulae and younger stars. Our own **Milky Way** is a swirling spiral galaxy.

An elliptical galaxy.

The central region of a spiral galaxy similar to our Milky Way.

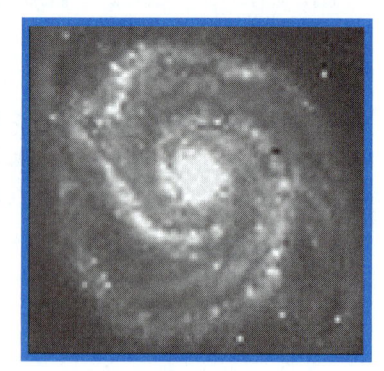

A spiral galaxy.

★ Other galaxies have **irregular shapes**. They could, for example, be in the form of clouds or rings.

Galaxies in turn are grouped into **clusters**. These clusters are giant bands numbering from thousands to millions of galaxies spread across space.

THE SOLAR SYSTEM

Earth, several other planets, asteroids, comets, and dust orbit the star known as the **sun**. Together, these bodies make up what is known as the **solar system**. The sun is the center of our solar system. It is the largest object in our solar system and contains more than 99% of the solar system's mass. The sun is so enormous in size that it could hold a total of 1.3 million planets the size of Earth.

THE PLANETS

Planets are objects of rock, metal, ice and gas that circle the sun. They do not give off their own light, as the sun does. There are eight planets in our solar system. They range from tiny rocky planets to huge gas giants with rings. The planets, in their order from the sun, are Mercury, Venus, Earth, Mars, Jupiter, Saturn, Uranus, and Neptune. The largest planet is Jupiter. It is so large that all of the other planets could fit inside it — it is ten times the width of Earth. In general, the farther away a planet is from the sun, the colder it is.

APPLYING WHAT YOU HAVE LEARNED

This table shows that Earth is 149.6 million kilometers from the sun, or 93 million miles away. Suppose there was an imaginary road from Earth to the sun. If you were in a car traveling 100 km an hour, it would take you 1,490,000 hours or just over 170 years to reach the sun. If it took 170 years to drive from Earth to the sun, how many years would it take to drive from Jupiter to the sun at the same speed?

DISTANCE OF PLANETS FROM THE SUN

Planet	Distance from the Sun
Mercury	57.9 million km
Venus	108.2 million km
Earth	149.6 million km
Mars	227.9 million km
Jupiter	778.3 million km
Saturn	1,427.0 million km
Uranus	2,871.0 million km
Neptune	4,497.1 million km

Answer: _____

THE ROLE OF GRAVITY

Gravity is a force of attraction between any two pieces of matter. The force of gravity influences the movements of bodies both on Earth and in space. Gravity increases as objects move closer together. Gravity also increases with the masses of the objects. Objects with larger masses will have a stronger gravitational attraction towards one another than smaller masses that are the same distance apart.

THE MOVEMENT OF THE PLANETS

The force of gravity explains the movement of Earth and the other planets around the sun, as well as the movement of the moon around Earth. At any moment, each planet would move in a straight line and fly off into space if the force of gravity from the sun did not bend its orbit. As a result, all of the planets revolve around the sun in **elliptical** (*oval-shaped*) orbits.

APPLYING WHAT YOU HAVE LEARNED

◆ How would the planets move if the sun did not exert any gravitational force?
   _____

   Explain your answer. _____
   _____

COMETS

Comets have been observed since ancient times. There are Chinese records of Halley's Comet that date back to 240 BC. Comets are sometimes called "dirty mudballs," because they are mainly made of rock, gas, ice and dust. Their orbits are much longer than those of the planets, and often bring the comet both closer to and farther away from the sun. As a comet approaches the sun, some of its ice turns to gas, creating a gaseous envelope known as a *coma*, and a

Halley's Comet will return to the inner solar system in the year 2061.

giant, glowing tail. This tail, made up of dust and gas, is often visible from Earth. Halley's Comet, for example, becomes visible from Earth once every 86 years.

ASTEROIDS AND DWARF PLANETS

Asteroid means "little star." Asteroids are irregular masses of rock and metal that orbit the sun. Some asteroids are mainly rock; others combine stone and iron or other metals. Thousands of asteroids are located in the "asteroid belt" between Mars and Jupiter. The force of gravity keeps these asteroids orbiting the sun in elliptical orbits, similar to the planets. A few asteroids, like Ceres and Eris, are so large they are called "dwarf planets." Pluto is now also considered a dwarf planet.

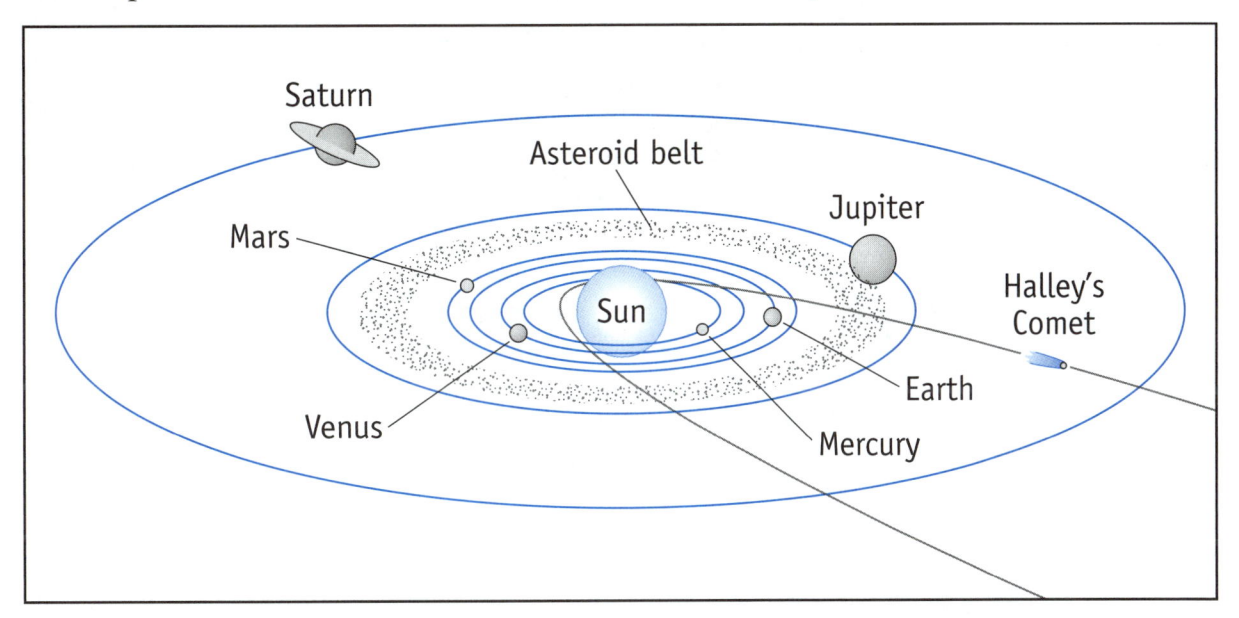

APPLYING WHAT YOU HAVE LEARNED

◆ How does the composition of an asteroid differ from that of a comet?

◆ How do the orbits of comets and asteroids compare to that of planet Earth?

METEORS AND METEORITES

If an asteroid or some other particle from outer space is pulled by gravity into a planet's atmosphere, it is called a **meteor**. Meteors are sometimes called "shooting stars," because they look like falling stars as they burn in Earth's atmosphere. A meteor that reaches the ground is known as a **meteorite**. Meteors mainly burn up in Earth's atmosphere, but if a large meteor struck Earth, it could cause tremendous damage.

Comets can also collide with a planet and cause meteor showers. In 1954, a comet fell into Jupiter, causing explosions much greater than those produced by nuclear bombs. An example of what happens when even a small meteor hits Earth can be seen near Winslow, Arizona. A crater was formed there about 50,000 years ago by an iron meteor only about 50 meters in diameter. The crater is 1,200 meters across and 200 meters deep.

Barringer Crater at Winslow, Arizona.

APPLYING WHAT YOU HAVE LEARNED

Scientists believe two giant asteroids have struck Earth. The impact of these crashes hurled hot rock and debris into space, which fell back to Earth and caused widespread fires. Smoke and dust from the crash and these fires blocked out light. The first asteroid struck 250 million years ago and killed more than 90% of Earth's ocean species and 70% of its land species. The second asteroid hit Earth 65 million years ago and probably caused the dinosaurs to become extinct.

◆ What do you think might happen if a large asteroid were to hit Earth today?

_____
_____

THE MOON

Gravity also affects the movement of the moon, which revolves around Earth in a circular orbit. The moon would otherwise travel through space in a straight line, but it is attracted towards Earth by the force of gravity. The interaction of these two forces causes the moon to orbit Earth.

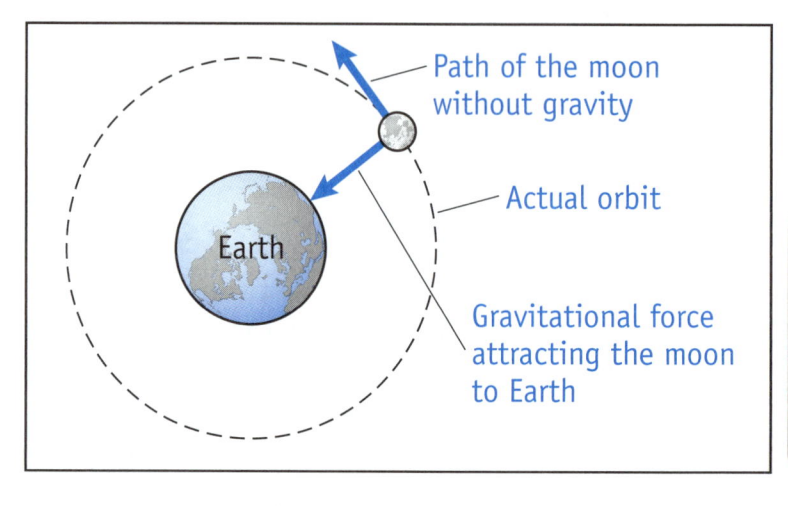

THE MOVEMENT OF EARTH

The planet Earth actually moves in two different ways at the same time: it **rotates** on its axis, and it **revolves** around the sun.

EARTH'S ROTATION

Earth **rotates**, or spins, around its axis — an imaginary line running through the center of Earth from the North Pole to the South Pole. This rotation takes 24 hours, causing day and night. Night occurs on those parts of Earth that are away from the sun's rays.

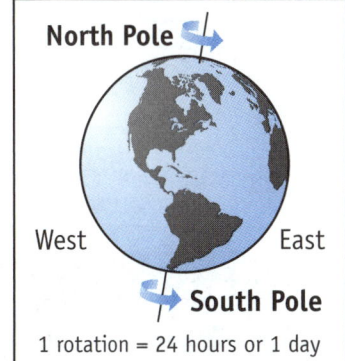

1 rotation = 24 hours or 1 day

EARTH'S REVOLUTION AND THE SEASONS

Earth revolves around the sun at the same time that it rotates on its axis. It takes just over 365 days (*one year*) for Earth to complete one revolution around the sun. Earth also **tilts** on its axis. Earth's axis tilts $23\frac{1}{2}°$ away from a line perpendicular to the line connecting Earth to the sun.

★ Because of this tilt, the sun's rays strike the Northern Hemisphere more directly in summer than in winter. The sun appears to rise higher in the sky, temperatures are warmer, and the days are longer.

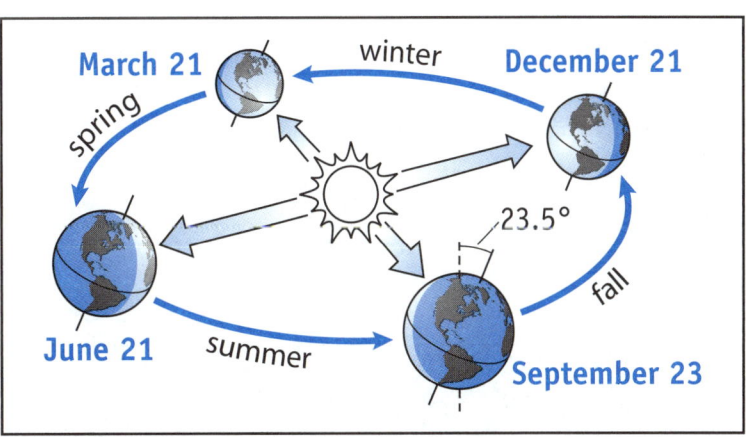

★ When it's summer in the Northern Hemisphere, it's winter in the Southern Hemisphere. Because the Southern Hemisphere is tilting away from the sun, it receives less direct solar rays. The area around the equator is not affected by Earth's tilt. It is always warm because it receives the sun's direct rays.

> The terms "day" and "year" can also be applied to other planets. A day on Venus is the time it takes Venus to rotate on its axis. A Martian year is the time it takes Mars to orbit the sun.

APPLYING WHAT YOU HAVE LEARNED

◆ When the Northern Hemisphere experiences winter, why is the Southern Hemisphere having summer? _____
_____
_____
_____

MOVEMENT OF THE MOON

The movements of Earth and the moon also explain the phases of the moon, eclipses, and tides.

PHASES OF THE MOON

The moon does not create its own light: it reflects the light of the sun. The amount of the moon we can see each night changes in a cycle that repeats itself each month. The moon begins as a thin sliver, grows into a crescent, expands into a full moon, becomes a crescent again, and finally narrows until it becomes completely dark in the night sky. These phases are caused by the moon's orbit and the different views from Earth of the sun's rays reflected by the moon.

PRINCIPAL PHASES OF THE MOON

New moon | First quarter | Full moon | Third quarter

★ When Earth and the sun are on the same side of the moon, the entire moon appears to be lit up at night, making a **full moon**.

★ When the moon is between Earth and the sun, a view from Earth cannot see any reflection of the sun's rays from the moon: a **new moon** is totally dark.

Because the moon rotates as it orbits Earth, the same side of the moon is always visible from Earth.

VIEWS OF THE MOON AT NIGHT

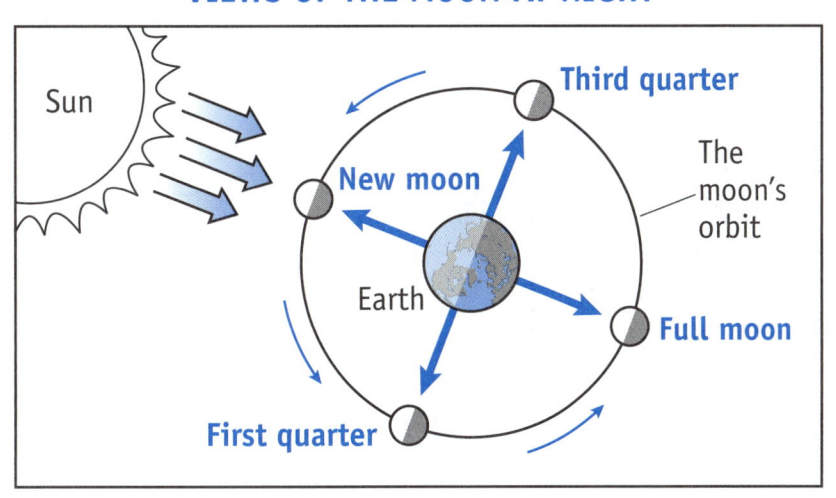

ECLIPSES

The movements of Earth and the moon are also responsible for eclipses. There are two kinds of eclipses.

- ★ **Solar Eclipse.** A **solar eclipse** occurs when the moon blocks sunlight from reaching Earth during the day. The sky darkens and only the sun's corona can be seen from behind the moon. It is dangerous to look directly at a solar eclipse even though the sun looks dark.

- ★ **Lunar Eclipse.** A **lunar eclipse** occurs when Earth's shadow blocks sunlight from the moon. A lunar eclipse only occurs during a full moon. If the moon passes into total shadow, the moon darkens; if it is in partial shadow, it turns yellow.

A total solar eclipse.

TIDES

Each day, the surface level of the oceans rises up and falls down during **high** and **low tide**. Tides are caused by the gravitational pull of the moon on Earth's ocean waters. The water of the ocean directly facing the moon is pulled by the moon's gravitational force.

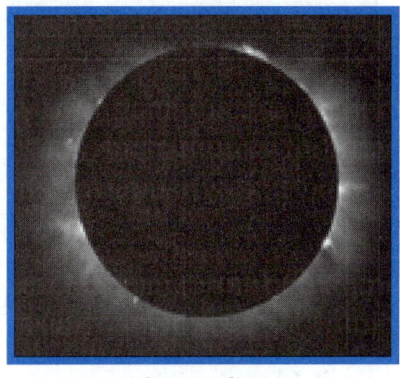

A lunar eclipse.

The water bulges towards the moon, creating high tide, a time when sea levels are at their highest. On the opposite side of Earth is another high tide, caused by the force of Earth's spin where the moon's pull is weakest. Sea levels become highest when the moon and sun are both lined up on the same side of Earth, and lowest when they are on opposite sides.

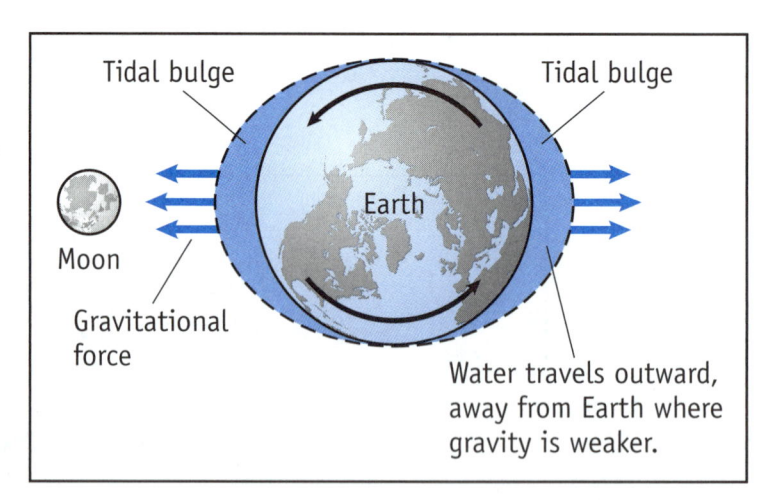

APPLYING WHAT YOU HAVE LEARNED

◆ Form a group with two other students to demonstrate an eclipse. Use a flashlight (or a table lamp) to represent the sun, a basketball to represent Earth, and a baseball to represent the moon. One student should hold the "sun" while a second, holding the basketball, circles the "sun." Finally, the third student, representing the moon, should circle "Earth." Move in a position where the moon blocks light from the flashlight when viewed from Earth. The places where the "moon" (baseball) casts a shadow on the "Earth" (basketball) represent those places where a solar eclipse would occur. You can use the same tools to demonstrate a lunar eclipse and to explain the tides. After you have completed your demonstration, write a summary of what you have done.

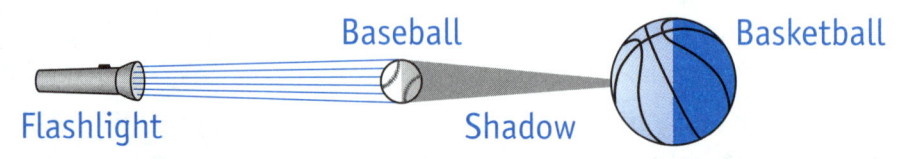

TOOLS FOR STUDYING SPACE

Because of the vastness of outer space and the difficulties it poses for survival, scientists generally cannot conduct laboratory experiments to explore it. Instead, they must depend upon their observations. As technology has improved, scientists have had more tools at their disposal to study space. For example, **telescopes** made with special lenses or mirrors make distant objects look large. Galileo was the first to use a telescope to study the night sky 400 years ago. He saw different planets more closely than ever before and discovered the moons of Jupiter.

Probing Outer Space. After the invention of rockets, scientists could actually send sophisticated instruments into outer space. Satellites and other space vehicles have now been sent into orbit. Unmanned space **probes** contain computers and cameras to take pictures of distant objects in space. These probes have been sent to Venus and Mars to study their atmospheres, to see if there is water, and to look for signs of life. Other space vehicles have placed **satellites** into orbit around Earth. These satellites allow continuous observation of outer space from beyond Earth's atmosphere.

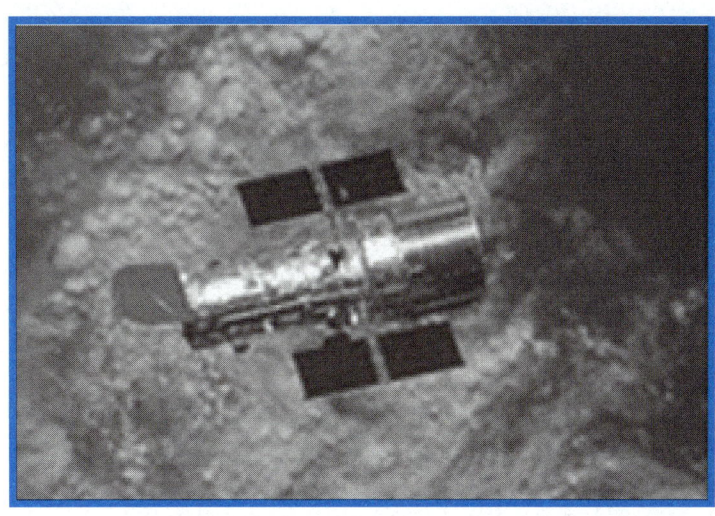

The Hubble Space Telescope orbits Earth, providing clear views of outer space.

Scientists are also able to study space with **manned spacecraft**. In 1969, American astronauts became the first humans to land on the moon. They took samples from the lunar surface to bring back to Earth for further study. Someday, astronauts may visit other planets or even journey to other stars.

APPLYING WHAT YOU HAVE LEARNED

◆ How does technology affect the ability of scientists to study the universe?
_____
_____

◆ The Hubble Space Telescope has shown scientists galaxies from over 12 billion light-years away. What do you think scientists can learn from these observations of outer space?
_____
_____

◆ Sending satellites, space probes, and manned space stations into outer space is very expensive. Do you think these expenses are justified? _____ Explain your answer. _____
_____
_____

WHAT YOU SHOULD KNOW

- [ ] You should know that the **universe** consists of galaxies, cosmic gases and dust, and empty space. A **galaxy** is a group of stars and other particles held together by gravity. Scientists classify galaxies by their shapes.
- [ ] You should know that **stars** are balls of superheated gases that produce energy through nuclear reactions. Each star passes through its own life cycle as its temperature and processes change.
- [ ] You should know that scientists measure distances between stars and galaxies in **light-years**.
- [ ] You should know that our **solar system** consists of the sun, its planets and their moons, dwarf planets, asteroids, comets, and other particles.
- [ ] You should know that objects in our solar system move in regular and predictable motions, which helps explain such events as years, day and night, the seasons, eclipses, tides and the phases of the moon. The force of **gravity** largely determines the motions of these bodies in space.
- [ ] You should know that scientists use a variety of tools to help them study space, including telescopes, probes, satellites, and manned spacecraft.

CHAPTER STUDY CARDS

Stars and Galaxies

★ **Stars.** Stars form out of clouds of gases and dust in space known as nebula. Stars produce energy through nuclear fusion, converting hydrogen into helium. Stars create all elements besides the lighter gases, as they go through their own life cycles.

★ **Galaxy.** A galaxy is a group of stars, gas and dust drawn together by gravitational force. Galaxies may be elliptical, spiral, or irregular.

★ **Light-Year.** The distance that light travels in one year.

Other Bodies in Our Solar System

★ **Comet.** A piece of rock, ice, and gas that circles the sun in a very long orbit. When it gets close to the sun, its gases and dust form a long, luminous tail. Halley's Comet is an example.

★ **Asteroid.** A piece of rock or metal orbiting around the sun in an elliptical orbit like the planets.

★ **Meteor.** An asteroid or particle that enters the atmosphere of Earth or another planet.

★ **Meteorite.** A meteorite is a meteor that has fallen to Earth before fully burning up.

CHAPTER 13: THE UNIVERSE

Gravity and Earth's Movement

★ **Gravity** is a force of attraction between any two objects. It increases with their masses.

★ **Effect of Earth's Movement.**
- Earth **rotates** on its axis, causing day and night.
- Earth **revolves** around the sun.
- The tilt of Earth's axis explains the change of seasons as Earth revolves around the sun.
- When it is winter in the Northern Hemisphere, it is summer in the Southern Hemisphere.

Effects of the Moon's Orbit

★ The moon does not cause its own light. It reflects the light of the sun.

★ **Lunar Phases.** The sun's reflected rays on the moon cause the different phases of the moon, from new moon to full moon.

★ **Eclipse.** Eclipses occur when the moon blocks the sun or Earth's shadow blocks sunlight from the moon.

★ **Tides.** Tides are the bulging of Earth's oceans, mainly caused by the moon's gravity.

CHECKING YOUR UNDERSTANDING

As the illustration below shows, when Earth is at its greatest distance from the sun, its Northern Hemisphere is tilted toward the sun.

1. What season takes place in the Southern Hemisphere at this time?

   A. spring
   B. winter
   C. autumn
   D. summer

   ES: A
   G8.2

   ◆ Examine the Question
   ◆ Recall What You Know
   ◆ Apply What You Know

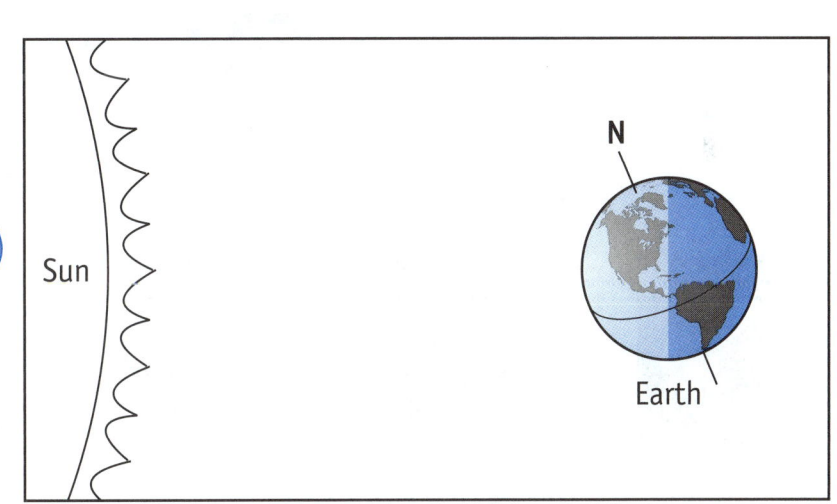

HINT: This question requires you to recall information about Earth's position in space and how it causes the seasons. You should recall that when the Northern Hemisphere tilts towards the sun, the Southern Hemisphere tilts away. What season does the Southern Hemisphere experience at this time? If you selected **Choice B**, you would be correct.

Now try answering some additional questions on your own:

2. Which list is in the correct order order from the smallest to the largest?
   A. galaxy → sun → Red Giant → dwarf star
   B. sun → dwarf star → Red Giant → galaxy
   C. sun → dwarf star → galaxy → Red Giant
   D. dwarf star → sun → Red Giant → galaxy

   ES: B
   G8.7

3. Which of these facts explains Earth's seasons?
   A. The Milky Way is expanding.
   B. Earth rotates around its axis every 24 hours.
   C. Earth tilts on its axis as it revolves around the sun.
   D. Earth's distance from the sun changes as it revolves.

   ES: A
   G8.1

4. Which process creates the phases of the moon observed from Earth?
   A. The moon reflects the sun's light.
   B. The moon acts as a nuclear reactor.
   C. The moon fuses hydrogen nuclei into helium.
   D. The moon casts a shadow on the sun's surface.

   ES: A
   G8.1

Use the illustration below to answer question 5.

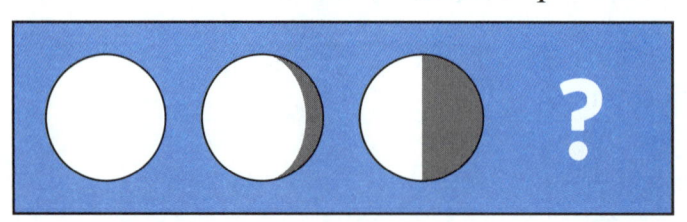

5. Which shows the next phase of the moon?

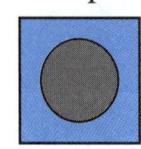

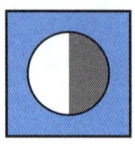

   A.   B.   C.   D.

   ES: A
   G8.1

6. What does a light-year measure?
   A. the distance light travels in one year
   B. the amount of energy light creates in one year
   C. the time it takes Earth to revolve around the sun
   D. the time it takes light to travel from the sun to Earth

   ES: B
   G8.6

   ♦ Examine the Question
   ♦ Recall What You Know
   ♦ Apply What You Know

Use the chart below to answer questions 7–9.

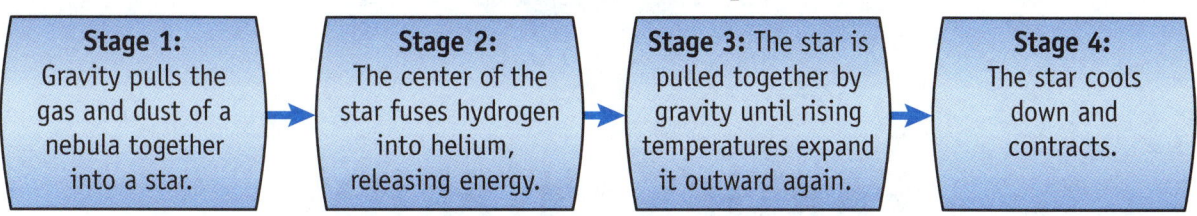

| Stage 1: Gravity pulls the gas and dust of a nebula together into a star. | → | Stage 2: The center of the star fuses hydrogen into helium, releasing energy. | → | Stage 3: The star is pulled together by gravity until rising temperatures expand it outward again. | → | Stage 4: The star cools down and contracts. |

7. During which stage can a star become a red giant?
   A. Stage 1
   B. Stage 2
   C. Stage 3
   D. Stage 4

8. During which stage will a star form a dwarf or black hole?
   A. Stage 1
   B. Stage 2
   C. Stage 3
   D. Stage 4

9. In which stage is our sun currently found?
   A. Stage 1
   B. Stage 2
   C. Stage 3
   D. Stage 4

Use the Venn diagram below to answer question 10.

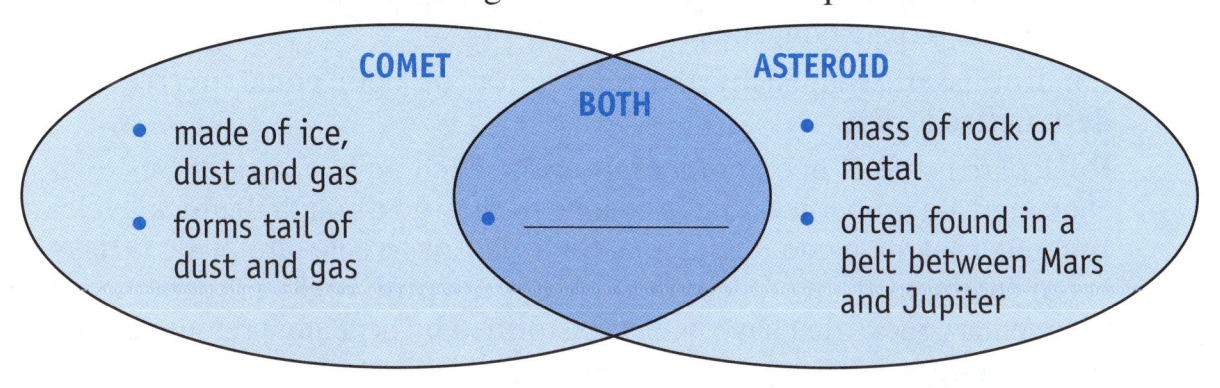

COMET
- made of ice, dust and gas
- forms tail of dust and gas

BOTH
- _____

ASTEROID
- mass of rock or metal
- often found in a belt between Mars and Jupiter

10. What is missing from the blank line in the Venn diagram?
    A. orbit the sun
    B. create a coma of gases
    C. produce their own light
    D. move closer to and farther from the sun than any planets

    ♦ Examine the Question
    ♦ Recall What You Know
    ♦ Apply What You Know

11. Scientists use many different tools to learn about stars, planets and other bodies in outer space.

    In your **Answer Document**, describe two tools used by scientists today to study outer space. (2 points)

CHAPTER 14

EARTH'S INTERIOR AND LANDFORMS

In this chapter, you will learn about Earth's interior and the shaping of its landforms.

— MAJOR IDEAS —

A. Earth's interior consists of an **inner** and **outer core**, a **mantle**, and a **crust**.

B. Earth's **crust** is divided into **tectonic plates** that ride on top of slow moving currents of **magma** (*hot, semi-molten rock*) in the mantle.

C. Most major geological events, such as earthquakes, volcanic eruptions, and mountain building, are caused by tectonic plate motion. Different types of plate boundaries (*divergent, convergent, and transform plate boundaries*) create typical landforms.

D. Landforms are shaped by the interaction of both **constructive** and **destructive** forces.

E. **Rocks** are made of one or more **minerals**. Each type of rock has its own characteristic properties. Rocks change as they go through the rock cycle. **Igneous rocks** are cooled magma. **Sedimentary rocks** are made of layers of sediment from eroded rocks. **Metamorphic rocks** are igneous or sedimentary rocks that have been transformed by heat and pressure below Earth's surface.

EARTH'S INTERIOR

Scientists use satellite images, globes, maps and models to analyze the size and shape of Earth and its interior. By studying earthquakes and volcanoes, scientists have concluded that Earth consists of a series of three distinct layers: **crust**, **mantle** and **core**.

CRUST

The **crust** forms a thick skin around Earth, much like the crust on a loaf of bread. All life takes place on this topmost layer. Earth's crust is made of solid rock. Oceanic crust forms the floor below the oceans. About 5 to 8 kilometers thick, oceanic crust is made of heavy, dense rock. The crust in land areas is much thicker (*between 30 and 50 km deep*) than oceanic crust, but it is made of lighter, less dense rock.

MANTLE

Below the crust is an area of hot, dense rock known as the **mantle**. Almost 3,000 km thick, the mantle makes up most of Earth's volume. The top of the mantle is solid, like the crust. As one goes deeper into Earth, both temperature and pressure rise. About 100 km below Earth's surface, the rock is near the melting point and becomes semi-solid or plastic.

CORE

The center of Earth is known as the **core**. It consists mainly of iron and nickel. Earth's core is extremely hot, with temperatures reaching well above 5,000° C. Radiation from radioactive

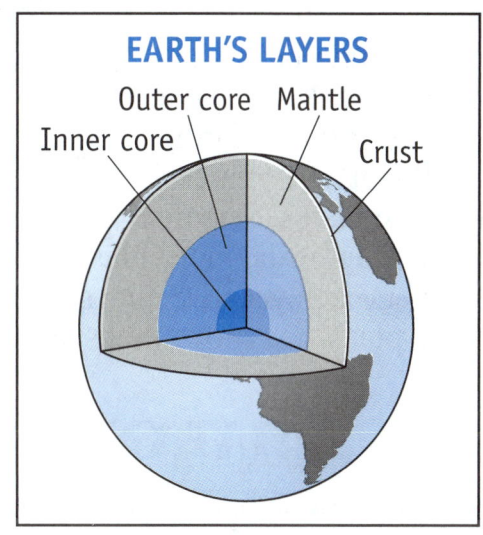

This model shows Earth's interior.

substances and thermal energy from Earth's formation create this heat, known as **geothermal energy**. The **outer core** is liquid. On the surface of Earth, the metals of the core would boil at these high temperatures, but pressure keeps the outer core in a liquid state. The **inner core** is made up mainly of iron. Although it is even hotter than the outer core, the tremendous pressure keeps the inner core solid. The movement of Earth's metallic core is the source of Earth's magnetic field.

PLATE TECTONIC MOTION

If you look at a map of the world, you may notice that different continents seem to fit together like a giant puzzle. For example, eastern South America seems to fill the space below West Africa. Mountain ranges that end at one coastline seem to continue again on the other coastline. Most scientists believe that several or even all of the present continents of the world once fit together into a single, giant continent. Gradually, this large land mass separated and its pieces drifted apart.

PLATE TECTONIC THEORY

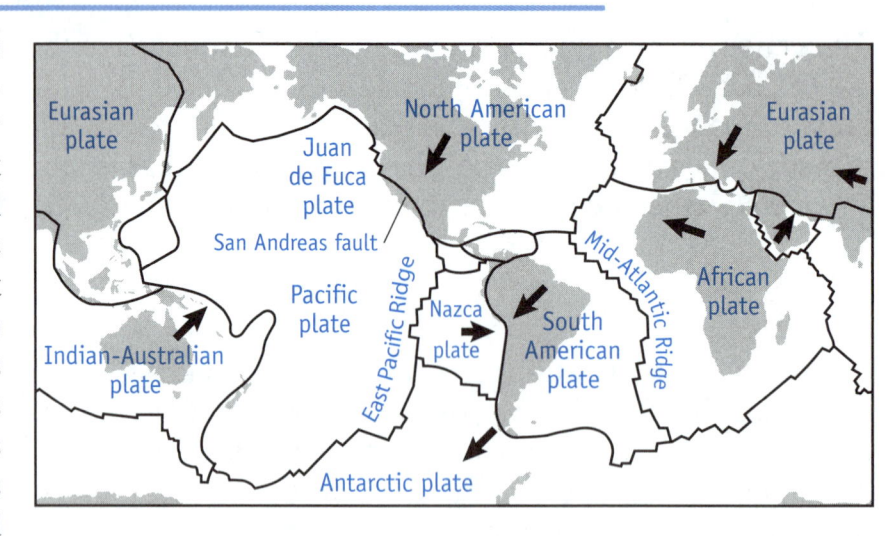

Scientists refer to these ideas as the **"plate tectonic" theory**. They identify Earth's crust and part of the solid upper mantle as the **lithosphere**. About 100 km (*or sixty miles*) thick, the lithosphere is divided into large slabs of solid rock known as **tectonic plates**. Earth's continents are attached to these plates. Scientists believe that the tectonic plates act like solid chunks floating on top of the more plastic part of the mantle. The plates move only a few centimeters each year. Despite such slow movement, over hundreds of millions of years these plates can move thousands of kilometers.

WHAT CAUSES PLATE MOVEMENT?

Scientists hypothesize that **thermal energy** and **gravity** may be responsible for the movement of tectonic plates.

THERMAL ENERGY

Inside the mantle, semi-solid rock is heated. As it is heated, it expands and becomes less dense. This lighter rock rises as gravity pulls down cooler, denser rock in its place. After the hotter rock rises and spreads, it begins to cool down. Once cooled, it sinks, creating a circular motion or current. This circular motion pushes the plates above.

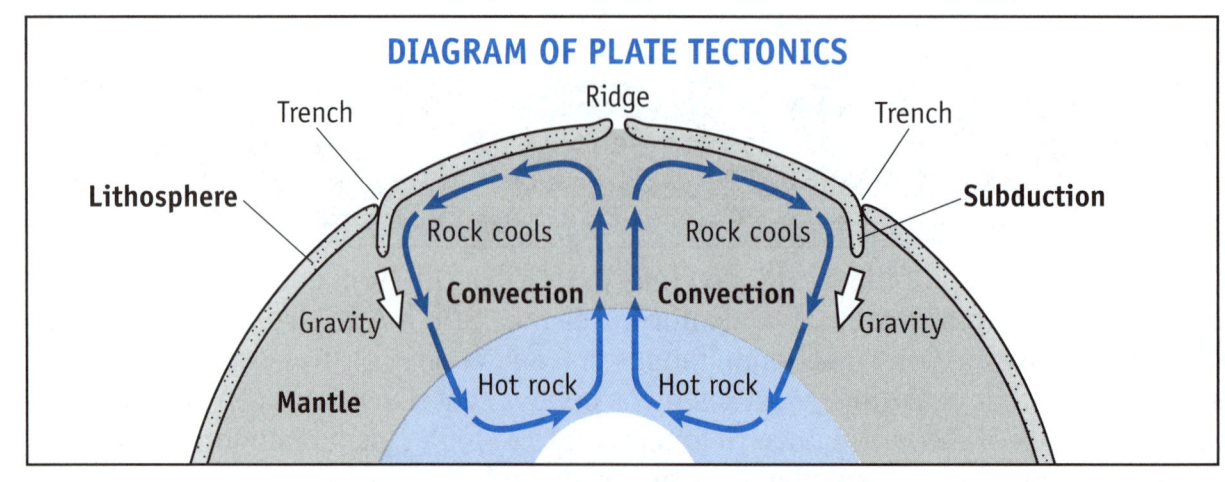

GRAVITY

Gravity also contributes to plate movement. You may recall that the force of gravity is greater when an object's mass is greater. When oceanic and land plates collide, the dense oceanic plate is pulled by gravity under the lighter land plate. As one end of the oceanic plate sinks, it pulls on the rest of the plate as well.

APPLYING WHAT YOU HAVE LEARNED

◆ Take a hard-boiled egg and crack its shell slightly. You will see the shell divide into several pieces. How are these pieces similar to and different from tectonic plates? _____
_____

◆ A scientist once said that "tectonic plate movement causes the surface of our planet to be in a state of constant change." What did she mean by this?
_____
_____

◆ Explain how gravity and thermal energy lead to the slow movement of tectonic plates. _____
_____

TYPES OF PLATE BOUNDARIES

Tectonic plates push and pull against each other like bumper cars in an amusement park. This results in a variety of different plate boundaries.

DIVERGENT PLATE BOUNDARIES

Divergent plate boundaries occur where two plates are moving apart, leaving a gap. Hot molten rock, known as **magma**, fills the gap between the plates, creating a mountainous ridge. As the plates move further apart, a **rift valley** is created between two ridges. There is an important divergent plate boundary in the middle of the Atlantic Ocean, which is adding new crust to Earth's surface. This is known as the **Mid-Atlantic Ridge**.

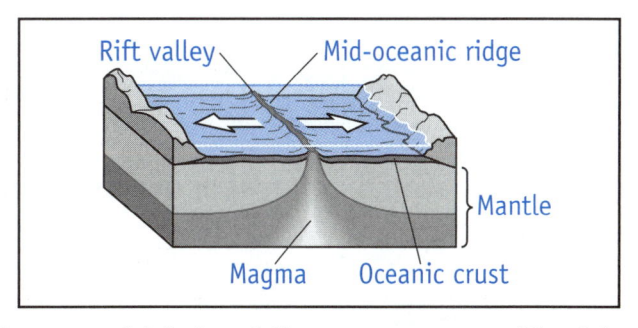

CONVERGENT PLATE BOUNDARIES

Convergent plate boundaries occur where two plates move together in slow collision. This leads either to folding or to one plate sliding under another. For example, the collision of the Indian Tectonic Plate with the Eurasian Tectonic Plate has led to the **folding** of Earth's crust and the creation of the Himalaya Mountains and Plateau of Tibet.

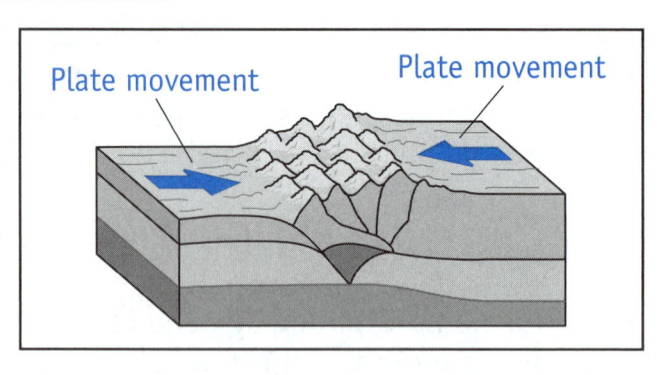

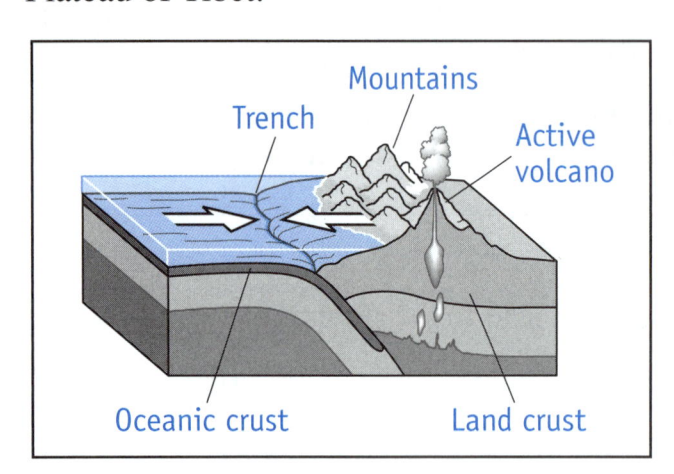

A special type of convergent plate boundary occurs where a dense oceanic plate collides with a lighter land plate. Gravity pulls the heavier ocean plate under the land mass and into the mantle. This process is occurring around the borders of the Pacific Ocean. Thus, as new crust is created in the middle of the Atlantic, older crust is being pulled back into the mantle along the edge of the Pacific.

TRANSFORM PLATE BOUNDARIES

A **transform plate boundary** occurs where one plate slides by another plate horizontally. The San Andreas fault in California is an example of a **transform plate boundary**, where the Pacific Plate slides past the North American Plate. Stress in transform plate boundaries often leads to earthquakes.

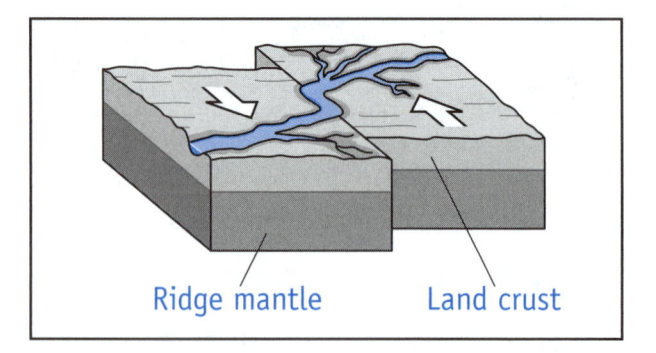

APPLYING WHAT YOU HAVE LEARNED

◆ How would you feel about living in a home along the San Andreas fault line? _____
_____

APPLYING WHAT YOU HAVE LEARNED

◆ Complete the table below by describing three types of plate boundaries:

Type	Description	Example
Divergent Plate Boundary		
Convergent Plate Boundary		
Transform Plate Boundary		

EFFECTS OF TECTONIC PLATE MOVEMENT

The movement of tectonic plates explains most major geological events and many of Earth's surface features. **Tectonic plate theory** helps us to understand how mountains are formed and why earthquakes and volcanoes occur where they do.

FOLDING AND MOUNTAIN BUILDING

When two plates converge, they often cause a folding of Earth's crust and build new mountains. As you learned on page 196, such folding has created the Himalaya Mountains and Plateau of Tibet — the highest landforms on Earth.

EARTHQUAKES

As plates move, they create tremendous stress in the rock at plate boundaries. Eventually, parts of Earth's crust may break, creating a **fault** and sending vibrations known as **seismic waves**. These vibrations cause earthquakes. When an earthquake occurs under or near the ocean, it creates immense ocean waves of destructive force known as **tsunamis**. In December 2004, a powerful underwater earthquake near the coast of Indonesia caused a devastating tsunami, leading to the deaths of thousands of people.

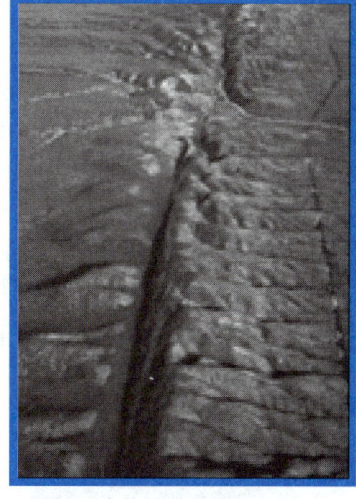

View of the San Andreas fault in Central California.

VOLCANOES

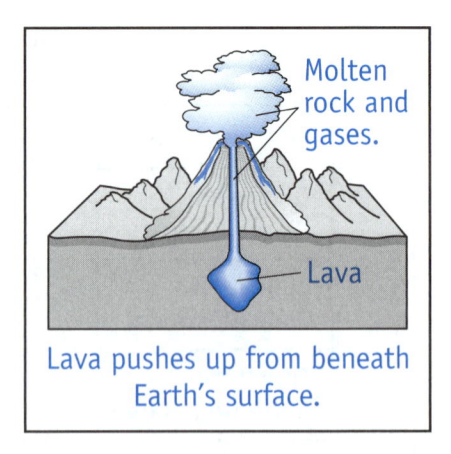

In places where tectonic plates diverge, or where one plate dives under another, pressure in Earth's mantle is reduced and some of the hot, solid rock turns into liquid. Any part of the tectonic plate that sinks into the mantle may also melt. Pockets of molten rock form beneath Earth's surface. This magma may break through weaknesses in Earth's crust. Magma, ashes and gases erupt and form a **volcano**. Once the magma reaches Earth's surface, it becomes **lava**. The location of most volcanoes and earthquakes is almost identical with the location of major plate boundaries. For example, the "Ring of Fire" around the Pacific Ocean — a zone of volcanoes and frequent earthquakes — coincides with the boundaries of the Pacific Tectonic Plate.

CONSTRUCTIVE AND DESTRUCTIVE FORCES

Earth's surface is undergoing constant change. Many of the processes caused by tectonic plate movement, like volcanoes and the folding of Earth's crust, serve to build up Earth's landforms. These are sometimes known as "**constructive forces**."

Other forces are at work slowly tearing Earth's landforms down. These are known as "**destructive forces**." The interaction of constructive and destructive forces is responsible for the various landforms we see on Earth's surface today.

WEATHERING

The wearing down of rocks on Earth's surface by the actions of wind, water, ice and living things is referred to as **weathering**. Water, for example, expands when it freezes. Cold nights and hot days often cause rocks to crack and break apart. Water may seep into cracks in rocks and expand these cracks if the water freezes. Rain and running water will also break down rock into smaller particles.

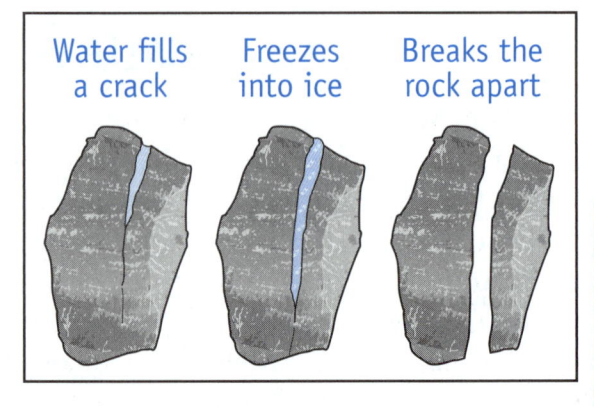

Running water can smooth rock, creating round stones and pebbles. **Chemical weathering** occurs when some chemicals **dissolve** rocks. Living organisms also cause weathering. Plants wedge their roots into the cracks of rocks, spreading them apart. Microscopic organisms may cause rocks to break down and disintegrate.

APPLYING WHAT YOU HAVE LEARNED

◆ Describe some of the different kinds of weathering that wear down rocks.
_____
_____

EROSION AND DEPOSITION

The process by which soil and rock are broken down and moved from one place to another is known as **erosion**. Once rock is broken into smaller particles, wind, running water, ice or gravity may cause this sediment to move to a new location. If you've ever been to a beach on a windy day, you can understand the power of sand blown by the wind. Rivers also carry and deposit sediment elsewhere. The action of ocean waves can wear down a rocky shoreline or cause the sand in a beach to move into the ocean, causing beach erosion. Erosion is a destructive force, but it leads to **deposition**, a constructive force. In places where sediment is deposited, Earth's surface is built up.

Erosion at the Grand Canyon.

GLACIERS

Glaciers are rivers of ice. They are formed in areas where there are very cold winters and cool summers. The snow that falls in the winter does not melt during the summer. Instead, it turns to ice. New snow then falls on top of this ice. As layers of snow build up, the snow's weight increases. This weight pushes on the ice and snow below, creating very thick, dense sheets of ice called **glaciers**. Glaciers move quite slowly. As they move, they scrape Earth's surface, picking up loose rock in their way, digging holes, wearing down mountains, and moving rocks and soil.

Moving glaciers can carve valleys through mountains. The rocks and boulders carried by the glaciers scrape the surface. A glacier can move millions of tons of material. Where it stops, a glacier will often leave behind rich soil. During the last Ice Age, glaciers covered much of what is now the northern United States.

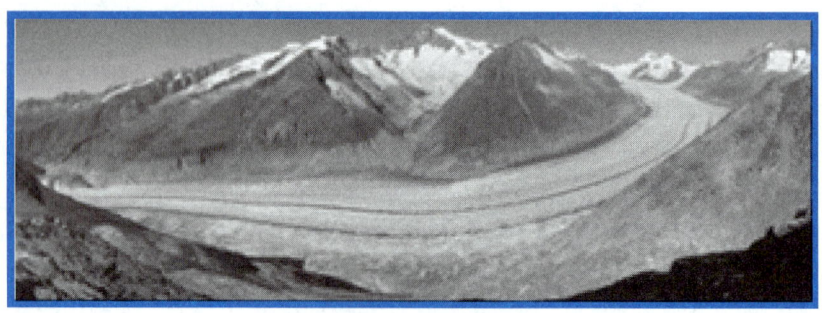

The path of a glacier.

APPLYING WHAT YOU HAVE LEARNED

★ What is erosion? _____
_____

★ How are glaciers formed? _____
_____

★ How do glaciers cause erosion? _____
_____

ROCKS

Much of Earth consists of what we call "rock." **Rock** is the main material found in Earth's crust and mantle. Each rock is made of one or more minerals. A **mineral** is any nonliving, solid substance in nature with a specific arrangement of atoms. Their atoms are arranged in symmetrical, geometric shapes known as **crystals**. There are thousands of different minerals, each with its own properties. Examples of minerals are diamonds, pyrite, quartz, feldspar and mica.

A diamond is a type of mineral.

TYPES OF ROCKS

There are many different kinds of rocks. Scientists classify rocks in a variety of ways — such as by their color, texture, and how they were formed. One common way to identify a rock is by the **minerals** it contains. Many rocks have crystals of quartz, mica, or feldspar.

Scientists also sometimes identify a rock by its hardness or **texture** — how the rock feels when you touch it. In 1812, Friedrich Mohs developed a mineral hardness scale now known as the **Mohs' Scale of Hardness** (see pg. 8). Many rocks also have different **layers**. The layers of a rock often help show how it was formed.

Geologists most often classify rocks by how they are formed. In general, there are three types of rocks:

★ **Igneous Rock.** Igneous rocks, sometimes called "fire rocks," were formed from magna or lava that has cooled. When the magma took a long time to cool, the mineral crystals in the rock are larger. Igneous rocks are usually heavy and they often have speckles, mineral crystals, grains, or veins. Granite is an example of this kind of rock.

An igneous rock.

★ **Sedimentary Rock.** When rock erodes, its particles are moved to a different place. Pieces of weathered rock and other materials become deposited on top of each other. This **sediment** may be pressed into layers. The top layers push down on the lower layers with their weight. Eventually, these layers of sediment turn into rock. Sedimentary rocks are usually lighter in color and weight than igneous rocks.

A sedimentary rock.

Their layers are often clearly visible. Because of the way sedimentary rocks are formed, the most recent layers are usually on top. However, because of the folding or faulting of Earth's crust, there may be shifts in layers. As a result, the youngest layer is not always found at the top. Fossils are sometimes found in sedimentary rock. Examples of sedimentary rock are limestone and sandstone.

★ **Metamorphic Rock.** The word *metamorphic* means "changed." Metamorphic rocks are formed by being changed or "morphed" into another kind of rock. A sedimentary or igneous rock may have moved back under Earth's surface. There, over millions of years, heat and pressure change it into a metamorphic rock, such as marble or slate.

A metamorphic rock.

All rocks can be classified as either igneous, sedimentary or metamorphic based on the way they were formed. Smaller rocks are created by the breakdown of these larger rocks through weathering and erosion.

THE ROCK CYCLE

As you have learned, igneous rock can be changed by erosion. Sedimentary rock can be changed by heat and pressure. In fact, much of the rock on Earth's surface is slowly recycled over time in what scientists call the **rock cycle**. Cooled magma forms **igneous rock**, such as granite or basalt. Weathering and erosion from water and air break down these rocks on Earth's surface into pebbles, sand and dust. These fragments pile up and become compressed into **sedimentary rock**, like sandstone.

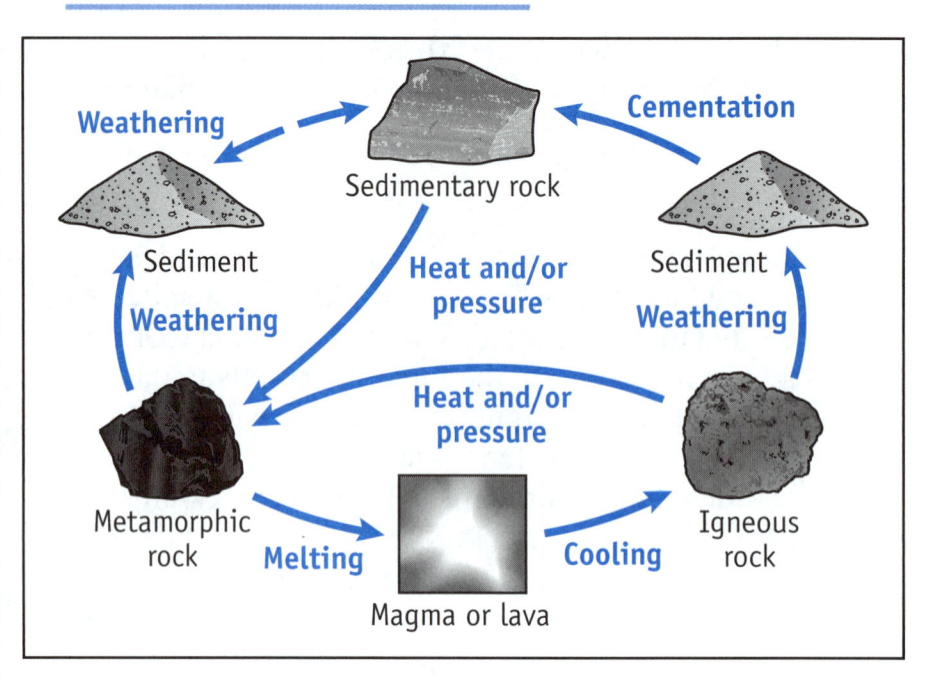

Changes caused by tectonic plate movement may bring sedimentary and igneous rocks below Earth's surface. Heat and pressure then change these rocks into **metamorphic rock**, such as marble or slate, or may even melt the rock completely, so that it forms new **igneous** rock.

APPLYING WHAT YOU HAVE LEARNED

Complete the following chart.

Rock	How Formed	Characteristics	Example
Igneous			
Sedimentary			
Metamorphic			

CHAPTER 14: EARTH'S INTERIOR AND LANDFORMS

WHAT YOU SHOULD KNOW

- [ ] You should know that Earth's interior consists of an inner and outer core, a mantle, and a crust.
- [ ] You should know that Earth's crust is divided into tectonic plates that ride on top of slow-moving currents of magma (*hot, semi-molten rock*) in the mantle.
- [ ] You should know that most major geological events, such as earthquakes, volcanic eruptions, and mountain building, are caused by tectonic plate motion. Different types of plate boundaries (*divergent, convergent, and transform plate boundaries*) create typical landforms.
- [ ] You should know that landforms are created through the interaction of both constructive and destructive forces. Destructive forces include weathering, erosion, and glaciers. Constructive forces include mountain building and deposition.
- [ ] You should know that rocks are made of one or more minerals. Each type of rock has its own characteristic properties. Rocks change as they go through the rock cycle. Igneous rocks are cooled magma. Sedimentary rocks are made of layers of sediment from eroded rocks. Metamorphic rocks are igneous or sedimentary rocks that have been transformed by heat and pressure below Earth's surface

CHAPTER STUDY CARDS

Earth's Interior

★ **Crust.** Outermost surface of Earth; oceanic crust is the floor beneath the oceans and is thinner and denser than the crust of land masses. The crust in land areas is much thicker than oceanic crust.

★ **Mantle.** Almost 3,000 km thick, the mantle is made up of hot, dense rock. As one moves deeper into Earth, the temperature and pressure of the mantle rises.

★ **Outer Core.** Molten nickel and iron.

★ **Inner Core.** Solid, mainly iron, at very high temperatures under pressure.

Tectonic Plate Movements

★ **Lithosphere.** Crust and top layer of mantle; divided into shifting tectonic plates.

★ **Plate Boundaries**
- **Divergent.** Two tectonic plates spread apart. Magma comes through the gap, creating a ridge.
- **Convergent.** Two tectonic plates come together. Oceanic crust dives under lighter land crust, or land plates fold into mountains.
- **Transform Plate Boundaries.** Two plates slide by horizontally, can cause earthquakes.

★ **Effects:** Earthquakes, volcanoes, folding of Earth's crust.

Rocks

★ **Rocks** are made of minerals found in Earth's mantle.

★ **Minerals** are nonliving, solid substances in nature with crystalline arrangements of atoms; e.g., quartz, diamonds.

★ **Types of Rocks:**
- **Igneous.** Made of cooled magma; has mineral crystals; e.g. granite.
- **Sedimentary.** Made of sediment (particles from eroded rock or living things); has visible layers; e.g., sandstone may have fossils.
- **Metamorphic.** Rock transformed by heat and pressure; e.g., marble.

Constructive & Destructive Forces

★ **Constructive Forces:**
- **Mountain-Building.** Folding of Earth's crust builds mountains.
- **Volcanoes.** Magma from interior erupts to the surface.
- **Deposition.** Sediment from erosion is deposited in new locations.

★ **Destructive Forces:**
- **Weathering.** Rock broken apart by wind, water, ice, and organisms.
- **Erosion.** Rock and particles are broken down and carried to a new location.
- **Glaciers.** Giant sheets of ice scrape Earth's surface.

CHECKING YOUR UNDERSTANDING

1. What do Earth's outer core and inner core have in common?
   A. Both are liquid.
   B. Both are solid.
   C. Both are mainly iron.
   D. Both are under the same pressure.

   ♦ Examine the Question
   ♦ Recall What You Know
   ♦ Apply What You Know

   ES: E
   G8.9

HINT: This question requires you to recall information about Earth's structure and processes. You should recall that the outer core is made up of extremely hot liquid. The inner core, although even hotter than the outer core, remains solid from the extreme pressure. Both, however, are mainly made of iron. Therefore, the correct answer is **Choice C**.

Now try answering some additional questions on your own:

2. Which best describes the density of Earth's crust?
   A. less dense than the core and mantle
   B. more dense than the core and mantle
   C. more dense than the core but less dense than the mantle
   D. less dense than the core but more dense than the mantle

   ES: E
   G8.9

CHAPTER 14: EARTH'S INTERIOR AND LANDFORMS   205

3. Which geologic event is most likely to occur at a divergent plate boundary?
   A. creation of rift valleys
   B. erosion of Earth's surface from scraping glaciers
   C. mountain building from a folding of Earth's crust
   D. deposition of sediment carried by fast-moving rivers

   ES: E
   G8.15

4. When the crust of a land mass collides with an oceanic plate, the land plate generally moves over the oceanic crust. What is the main reason the land plate moves on top of the oceanic plate?
   A. The crust of a land mass is less dense.
   B. The crust of a land mass deforms less easily.
   C. The crust of a land mass contains more minerals.
   D. The crust of a land mass melts at higher temperatures.

   ES: E
   G8.9

5. The diagram to the right illustrates an important scientific theory. What does this theory hypothesize?
   A. Earth moves around the sun.
   B. Earth's continents are slowly moving together.
   C. There is little difference in climatic patterns around the world.
   D. Millions of years ago, Earth's continents once formed one giant continent.

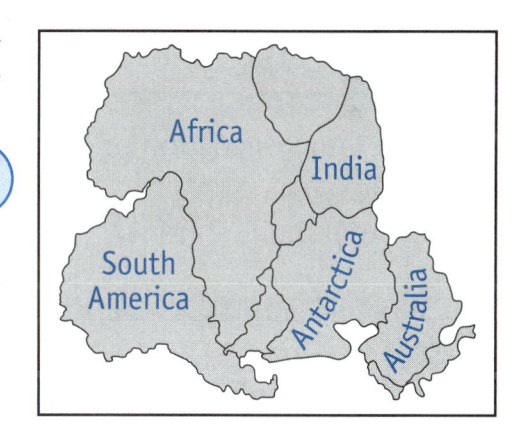

   ES: E
   G8.9

6. Why should the rock shown to the right be classified as a metamorphic rock?
   A. It has curved edges.
   B. It has a mixture of minerals.
   C. It was formed from other rocks.
   D. It has crystals from precipitation.

   ES: D
   G6.1

7. When granite is transformed by heat and pressure without entirely melting, what does it become?
   A. igneous rock
   B. sand and gravel
   C. sedimentary rock
   D. metamorphic rock

   ♦ Examine the Question
   ♦ Recall What You Know
   ♦ Apply What You Know

   ES: D
   G6.1

Use the information in the diagram below to answer questions 8 to 10.

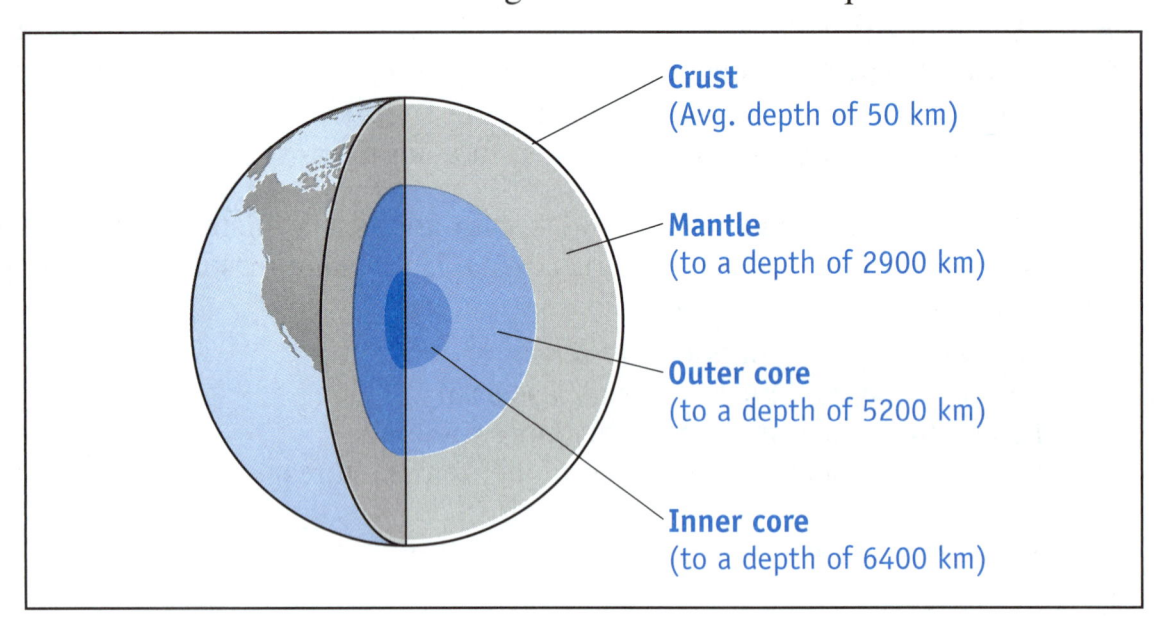

- Crust (Avg. depth of 50 km)
- Mantle (to a depth of 2900 km)
- Outer core (to a depth of 5200 km)
- Inner core (to a depth of 6400 km)

8. The thinnest section of Earth's crust can be found beneath what part of Earth?
   A. oceans
   B. mountain regions
   C. desert regions
   D. coastal plains

9. In which part of Earth do scientists believe temperatures reach over 5000° C?
   A. core
   B. crust
   C. mantle
   D. ocean floor

   ♦ Examine the Question
   ♦ Recall What You Know
   ♦ Apply What You Know

10. Scientists believe that Earth's interior can be divided into the layers shown in the diagram above. Which evidence best supports this hypothesis?
    A. data from oil deposits
    B. records from cyclone activity
    C. evidence from earthquakes and volcanoes
    D. signs from the exploration of other planets

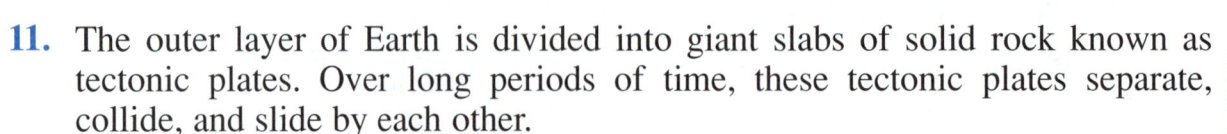

11. The outer layer of Earth is divided into giant slabs of solid rock known as tectonic plates. Over long periods of time, these tectonic plates separate, collide, and slide by each other.

    In your **Answer Document**, identify two effects of tectonic plate movement.

    Then explain how tectonic plate movement causes each of the effects you identified. (4 points)

CHAPTER 15

THE INTERACTION OF EARTH'S SYSTEMS

In the last chapter, you learned about the lithosphere — Earth's crust and land masses. In this chapter, you will learn about some of Earth's other systems.

— MAJOR IDEAS —

A. Earth's **lithosphere** consists of Earth's crust and land areas. Earth's **hydrosphere** consists of all the water on Earth. Earth's **atmosphere** is an envelope of gases surrounding Earth. These systems often interact.

B. The **water cycle** moves water and energy between the hydrosphere and atmosphere. The water cycle can lead to floods, tornadoes, and hurricanes.

C. **Pollution** affects the ability of Earth to absorb and recycle materials. Pollution also affects the ability of Earth's freshwater systems to support life.

D. Interactions between Earth's hydrosphere, atmosphere, and lithosphere lead to typical **weather patterns**. For example, different cloud types are often associated with particular weather conditions.

E. Scientists make observations and take measurements about the weather to create **weather maps**. Weather maps can be used to interpret the weather.

F. Particular temperature and precipitation patterns create specific climatic zones and **biomes**.

THE EARTH'S HYDROSPHERE

More than 70 percent of Earth's surface is covered by water. Scientists refer to this as the **hydrosphere**. About 97 percent of this water is in the ocean; most of the rest is frozen in the polar ice caps; less than one percent is actually found in the atmosphere, groundwater, and freshwater lakes and rivers.

THE WATER CYCLE

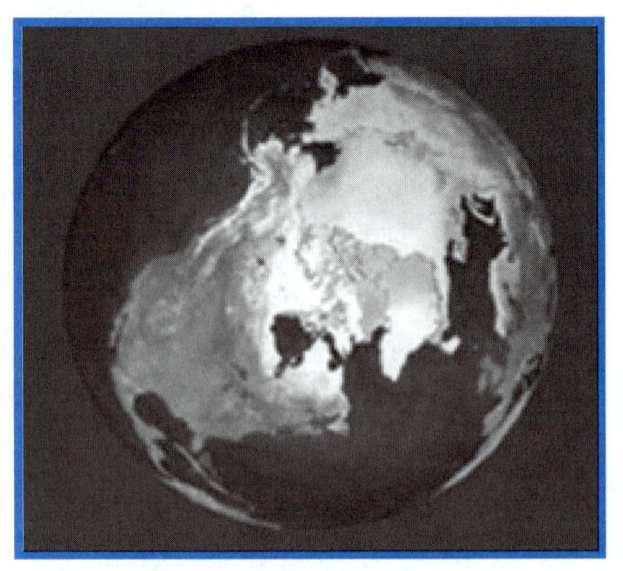

Just as Earth's lithosphere undergoes various processes, so does the hydrosphere. One of the most important is the **water cycle**. The water cycle begins when solar energy, radiating from the sun, heats the surface of the oceans. This transfer of energy causes some of the surface water to evaporate into the atmosphere, becoming **water vapor**, an invisible gas. Plants also create water vapor through **transpiration**, and animals do this through **perspiration**.

Earth's oceans hold most of its water.

The water vapor next **condenses** around dust and other particles into droplets small enough to float in the atmosphere as **clouds**. When the droplets join together to grow larger and heavier, they fall back to Earth's surface as **precipitation** — rain, snow or hail. Some precipitation returns to the ocean, but some falls on land where it is either absorbed by the ground or forms lakes, streams and rivers. Heavy precipitation can lead to flooding. Water absorbed by the ground sinks until it hits dense rock, where it collects as **groundwater**. Some of the surface water evaporates, but most of it will eventually drain back into the ocean.

THE WATER CYCLE

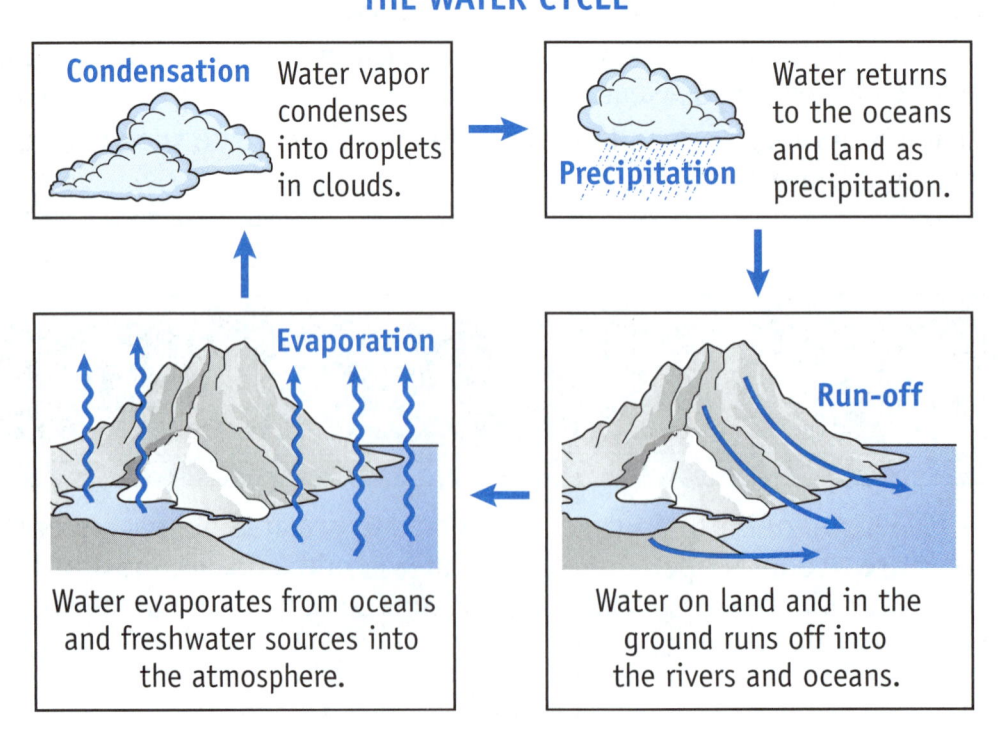

Condensation — Water vapor condenses into droplets in clouds.

Precipitation — Water returns to the oceans and land as precipitation.

Evaporation — Water evaporates from oceans and freshwater sources into the atmosphere.

Run-off — Water on land and in the ground runs off into the rivers and oceans.

There is significant interaction between matter and energy in the water cycle. Energy is required for water to evaporate. Because energy is absorbed by the water molecules as they turn into a gaseous state, the process of **evaporation** cools surrounding surface areas and the atmosphere. When water condenses into droplets, it releases thermal energy back into the air.

APPLYING WHAT YOU HAVE LEARNED

◆ Explain how the water cycle affects Earth. _____
_____
_____

POLLUTION THREATENS EARTH'S CYCLES

Other matter besides water is also transferred back and forth between Earth's surface, hydrosphere and atmosphere. For example, animals breathe out carbon dioxide into the atmosphere. Plants then use this carbon dioxide for photosynthesis. Decomposers break up the bodies of dead plants and animals, returning chemicals to the soil. Plants then absorb some of these chemicals through their roots. This recycling of materials through Earth's systems takes time. The ability of Earth to absorb and recycle materials through these **biogeochemical cycles** has been challenged by the growing pollution caused by human activities.

THE DANGERS OF POLLUTION

The rise of industry and the growth of the human population in the past two centuries has led to a decline in Earth's air and water quality. Car exhaust, smoke from factories, **smog** (*a mixture of fog and smoke*), and sewage from manufacturing and urban centers now pollute the air, freshwater sources, and groundwater. Oil spills cover some parts of the shoreline. Earth's systems cannot decompose some synthetic materials, such as certain plastics. Even those materials that Earth's systems can recycle, like carbon dioxide, are being produced far too rapidly for Earth's natural processes to absorb and process.

★ **Global Warming.** The burning of fossil fuels like coal and oil has greatly increased carbon dioxide in the atmosphere. Carbon dioxide and water act together to wrap Earth in a blanket, holding in heat. With more carbon dioxide in the air, less heat is able to escape, leading to a "**greenhouse effect**." Warmer temperatures have led to less precipitation and droughts in some areas. World leaders are becoming more aware of these global dangers.

★ **Pollution and the Depletion of Water Supplies.** The availability of fresh water is essential to humans and many other organisms. The dumping of wastes by manufacturers and growing cities threatens our drinking water and disrupts the ecological balance in rivers and oceans. The use of fresh water by growing cities and droughts caused by global warming have led to the **depletion** of many sources of water. Lakes and rivers have become less suitable for many species of plants and animals; some have even dried up altogether.

APPLYING WHAT YOU HAVE LEARNED

◆ Pollution is a growing problem for all living organisms on Earth. What steps should world leaders take to reduce global pollution? _____
_____

◆ What steps are being taken in your own community to reduce pollution?
_____
_____

THE ATMOSPHERE AND WEATHER

The **atmosphere** refers to the envelope of gases surrounding Earth. Our atmosphere is mainly made up of nitrogen (78%) and oxygen (20%) gases. The atmosphere absorbs solar radiation, moderates temperatures, and distributes water. The **weather** refers to the condition of the atmosphere. To describe it, scientists use:

★ **Temperature.** Scientists measure how hot or cold the air is in degrees (F or C).

★ **Precipitation.** Scientists measure the fall of rain, snow, hail, or sleet in inches or centimeters.

★ **Wind.** Scientists measure the speed and direction of the wind.

★ **Barometric Pressure.** The force with which the weight of air pushes on the ground is known as **air pressure**. Scientists use a **barometer** to measure air pressure. They see if the barometer is rising or falling to predict what kind of weather lies ahead. Cold or moist air is usually denser than warm or dry air. Falling pressure indicates probable rain.

TYPICAL WEATHER PATTERNS

Weather patterns result from the interaction of Earth's systems — landforms, heating by the sun, the spinning of Earth, and water from Earth's oceans and lakes.

★ Temperatures become warmer the closer one gets to the equator — the imaginary line around Earth's middle. **Hurricanes** occur in tropical areas when warm ocean water evaporates so quickly it spirals upwards. Hurricanes are accompanied by strong winds and heavy rains. In the middle of the U.S. when dry, cool air meets a mass of warm air, the warm air may rise quickly, causing a **tornado** — a destructive, funnel-shaped rotating column of wind.

★ Because air cools as it rises over a mountain, the ocean side of a mountain often has heavy rainfall. The air loses its moisture and becomes drier by the time it reaches the other side of the mountain.

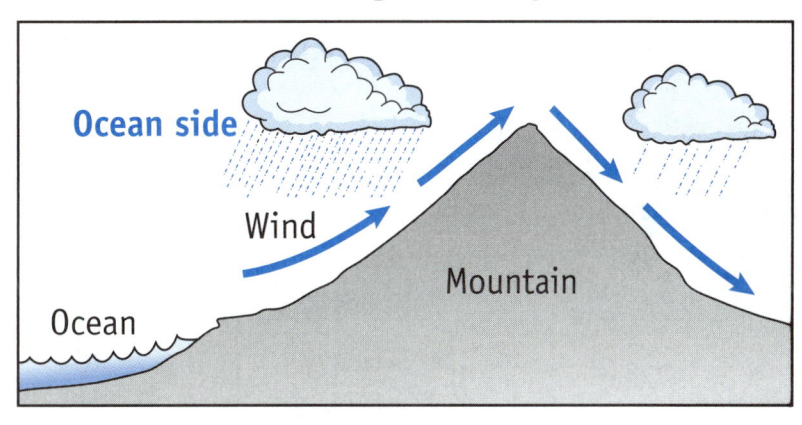

★ Differences between land and water also affect climate. Water requires more energy to change its temperature. Lakes and oceans stay cooler than land in summer and warmer in winter. This affects air flowing over these areas.

CLOUDS

Clouds are formed by ice and water droplets that have condensed in the atmosphere. Cloud formations are usually used to help predict the weather.

★ **Cirrus Clouds** form out of ice crystals high in the sky. Cirrus clouds look feathery, thin and white. They often indicate good weather.

★ **Cumulus Clouds** appear as puffy white clouds with flat bottoms. They usually indicate good weather.

★ **Stratus Clouds** form low in the sky, appearing as white blankets or layers that can cover the entire sky. They often indicate a gray, dull day with rain.

★ **Cumulonimbus Clouds** appear as white blankets or mushrooms that reach far upwards. They often bring heavy rain, thunderstorms, and lightning.

Cirrus Clouds.

Stratus Clouds.

APPLYING WHAT YOU HAVE LEARNED

✦ Look out the window of your classroom. What kind of cloud formations do you see? _____

✦ What future weather do these clouds indicate? _____
_____

RECORDING WEATHER INFORMATION

Scientists often record weather information. They record the temperature, the direction and speed of the wind, barometric pressure, and precipitation (*rain, snow, and hail*). Scientists use special weather station symbols to record this information:

Cloud Cover	Precipitation	Weather Station Information
○ Clear ● Overcast ◐ Partly cloudy	⋮ Heavy rain ✻✻ Heavy snow ═ Fog	Wind direction (to the north), Wind speed, Air temperature (° C) (It may also be recorded in ° F.), 18 | 1029, Air pressure, Pressure change, Cloud cover

WEATHER MAPS

A **weather map** is a special kind of map showing weather patterns. For example, it can show temperatures, air pressure, cloud conditions, and precipitation using the symbols above. A weather map can also be used to indicate the location of large bodies of cold or warm air, or high or low pressure areas. In this map, there is a high pressure area to the west of Ohio. This area also has a cold front, which is moving eastwards towards Ohio. There are very few clouds over Columbus, where the temperature is 60° F and the winds are moving south.

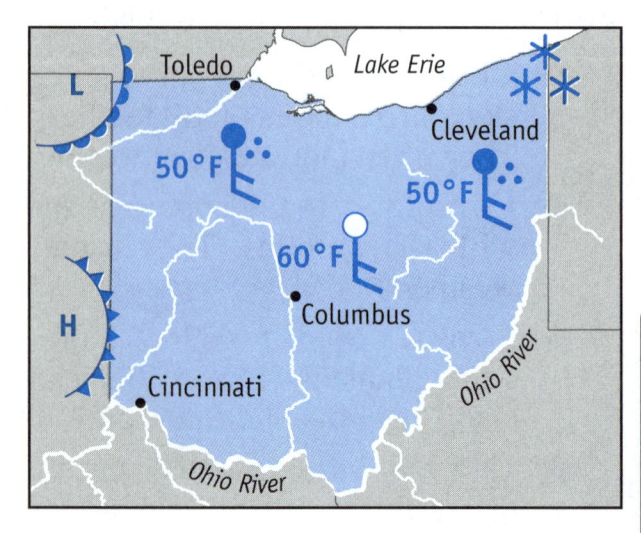

AIR PRESSURE

Although the molecules of gases in the air are so small they are invisible to the eye, they have weight and take up space. Air pressure is the force with which these molecules press downward towards the ground.

★ **High Pressure.** When air flow is blocked, it can cause a mass of air to build over a particular area. This causes the air pressure to increase. On a weather map, areas of dense air exert **high pressure** and signal the approach or continuation of fair weather. These areas are generally marked with "H."

★ **Low Pressure.** When air is heated, it rises. This lowers air pressure at the surface. Areas on a weather map, marked with an "L," indicate **low pressure**. In general, falling pressure usually signals the onset of stormy weather.

WEATHER FRONTS

A **front** occurs whenever cold and warm air masses meet. Specific weather conditions are associated with each kind of front. When fronts move into an area, they often bring precipitation, and changes in wind direction and temperature.

Type of Front	General Weather Conditions	Cloud Conditions
Warm Front	These usually move slowly. A warm front will move more slowly over a mass of cold air, bringing slow, steady rain and warmer temperatures.	Warm fronts are associated with mid-level clouds, such as stratus or altocumulus.
Cold Front	These are usually fast moving. Often they move under a mass of warm air and push the warm air upwards. This can lead to thunderstorms, high winds and even tornadoes.	Quickly rising warm air, pushed up by the cold front, creates cumulonimbus clouds.
Stationary Front	This occurs when two air masses meet, but neither pushes out the other. This often leads to rain and slightly warmer temperatures.	Stratus and cirrus clouds are associated with stationary fronts.
Occluded Front	This front occurs when warm and cool air masses meet and mix. This can bring slow, steady rain, like a warm front, and storms, like a cold front.	These have mid-level clouds, like a warm front, and cumulonimbus clouds, like a cold front.

> **APPLYING WHAT YOU HAVE LEARNED**
>
> ◆ Listen to the weather report on television. Then in your notebook create a weather map of your area indicating the type of weather you will be having over the next few days. Be sure to indicate precipitation, high and low pressure areas, and any warm or cold fronts that might be approaching.

CLIMATE AND BIOMES

The **biosphere** refers to all life on Earth. Weather patterns and climate influence what kinds of life forms can successfully live in each particular location. Based on the interaction of climate, natural resources and life forms, scientists have identified several different **biomes**, or geographic regions with certain types of life.

TEMPERATE FOREST

Temperate forests develop in regions where there is ample rain and moderate temperatures with cool winters. Trees change colors in fall and lose their leaves in winter. Temperate forests have a wide range of plant and animal life.

TROPICAL RAINFORESTS

Tropical rainforests develop in tropical areas near the equator where there is ample rainfall and warm temperatures year-round. Large trees cover the area with their leaves, forming a **canopy**. Tropical rainforests have a great abundance of animal and plant life with greater biological diversity than any other biome.

GRASSLANDS

Grassland areas exist where the climate is drier and there is not enough rainfall to support large numbers of trees. Instead, grasses dominate. Grassland areas often support large grazing animals, like antelope or bison.

Cattle are found mainly in grassland areas.

DESERTS

Deserts are regions with very little rainfall. Deserts, such as the Sahara Desert in Africa, have their own special forms of plant and animal life. These organisms can often store water for long periods.

ALPINE

Alpine biomes are found in high mountain areas. Because of their high altitudes, they have cold, snowy winters and short, cool summers. There are fewer forms of plants in alpine biomes because of less carbon dioxide and colder temperatures. Night temperatures are always below freezing.

TUNDRA

Tundra biomes are found closer to the polar regions. The soil is so cold that trees cannot grow there. Much of the ground is frozen for part of the year. Grasses and small shrubs are able to adapt, while mammals and birds migrate to these regions in the warmer spring and summer months.

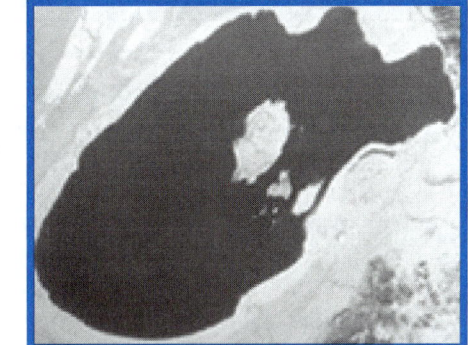

A frozen tundra.

APPLYING WHAT YOU HAVE LEARNED

◆ Identify the biome where you live and describe its characteristics.

___

WHAT YOU SHOULD KNOW

☐ You should know that Earth's **lithosphere** consists of its crust and land areas. Earth's **hydrosphere** consists of all its water. Earth's **atmosphere** is an envelope of gases surrounding Earth.

☐ You should know that the **water cycle** moves water and energy between the hydrosphere and atmosphere. Events in the water cycle can lead to such actions as floods, tornadoes, and hurricanes.

☐ You should know that **pollution** can affect the ability of Earth's natural processes to absorb and recycle materials. Pollution reduces the ability of rivers, lakes and groundwater to support life.

☐ You should know that interactions between Earth's hydrosphere, atmosphere, and lithosphere lead to typical **weather patterns**. For example, different **cloud types** and **fronts** are often associated with particular weather conditions.

☐ You should know that scientists make observations and take measurements about the weather to create **weather maps**. Weather maps can be used to interpret the weather.

☐ You should know that particular temperature and precipitation patterns create specific climatic zones and **biomes**.

CHAPTER STUDY CARDS

Hydrosphere
★ The **hydrosphere** is made up of all water on Earth's surface.
★ Fresh water is needed for life.
★ **The Water Cycle.** Water circulates through evaporation, condensation, precipitation, and run-off.
★ **Other Cycles.** Other cycles transfer materials, such as carbon dioxide or nutrients, between the lithosphere, hydrosphere, and atmosphere.
★ **Pollution.** Earth's capacity to absorb and recycle materials requires time. Pollution (like smoke, smog, or sewage) threatens these processes.

Atmosphere
★ The **atmosphere** is an envelope of gases around Earth. It is mainly made up of nitrogen and oxygen.
★ These gases absorb solar radiation, moderate temperatures and distribute water.
★ The atmosphere creates distinct weather patterns. Heating of the atmosphere and Earth's spin create wind patterns. Surface features like mountains also affect weather.
★ **Cloud Types.** These help predict weather.
★ **Weather Maps.** Help to interpret weather.
★ **Biomes.** Variations in climate lead to Earth's different **biomes** — such as temperate forests, grasslands and alpine areas.

CHECKING YOUR UNDERSTANDING

1. A student looks outside her window after it has rained and sees puddles of water. Later that same day, the puddles have disappeared. What process explains why the puddles have disappeared?

   A. run-off
   B. evaporation
   C. precipitation
   D. condensation

   ES: C
   G7.3

> **HINT:** To answer this question, you need to know the different steps of the water cycle. When water turns from a surface liquid water to water vapor in the air, it evaporates. Therefore, the correct answer is **Choice B**.

Now try answering some additional questions on your own:

2. Which symbol is used to represent a moving cold front on a weather map?

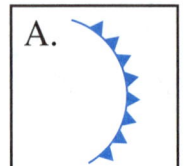

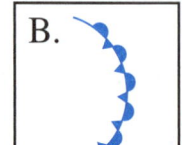

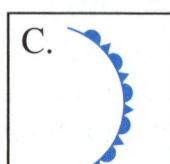

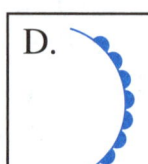

   A.   B.   C.   D.

   ES: C
   G7.7

CHAPTER 15: THE INTERACTION OF EARTH'S SYSTEMS   217

3. In which part of the diagram on the right is water changing from a liquid to a gas?
   A. A
   B. B
   C. C
   D. D

WATER CYCLE

4. When is air pressure usually highest?
   A. when it is cold and dry
   B. when it is warm and dry
   C. when it is cold and humid
   D. when it is warm and humid

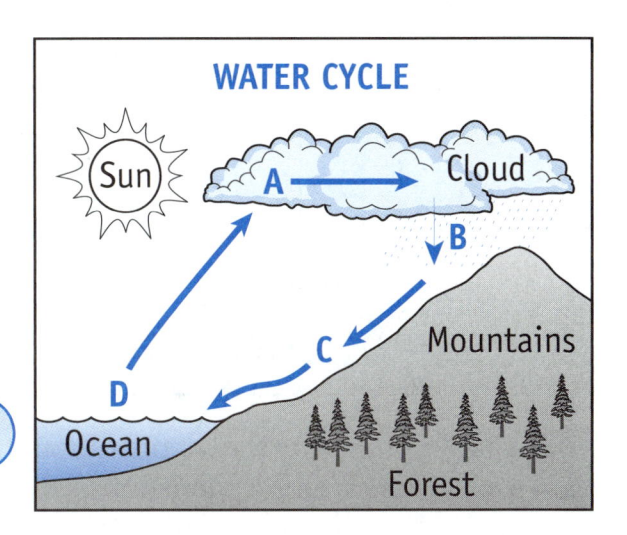

Use the following information to answer questions 5 to 7.

Lake Chad, just south of the Sahara Desert, was one of Africa's largest bodies of fresh water. It provided water to more than 20 million people in the four countries surrounding the lake. Fifty years ago, the lake was 26,000 km² — close in surface area to Lake Erie. Today, the lake has shrunk to less than 1,500 km².

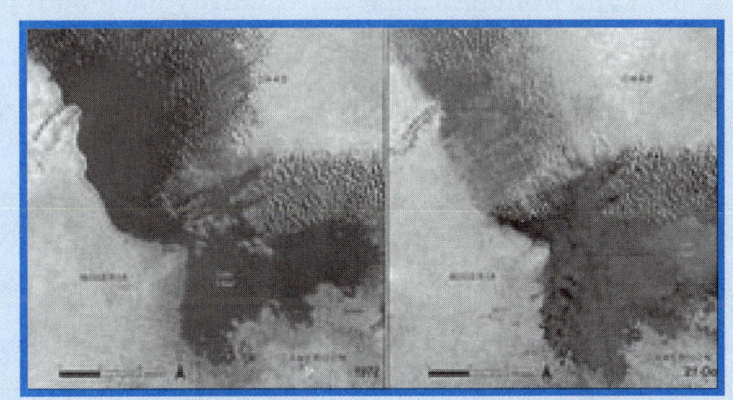

Lake Chad 1972   Lake Chad 2001

5. What is the most likely cause of this change to Lake Chad?
   A. Frequent rains have caused flooding.
   B. Soil erosion has allowed the lake to drain.
   C. Tornadoes have deposited sand in the lake.
   D. Global warming has led to repeated droughts.

6. What has been one important effect of the shrinkage of Lake Chad?
   A. Larger fish have eaten most of the plant life in the lake.
   B. There are fewer fish in the lake to feed local inhabitants.
   C. Plants in the lake have benefited from increased sunlight.
   D. Reductions in salt concentrations have killed some lake species.

7. What process does the shrinkage of Lake Chad best illustrate?
   A. pollution
   B. depletion
   C. condensation
   D. transpiration

AIR POLLUTION IN A CITY

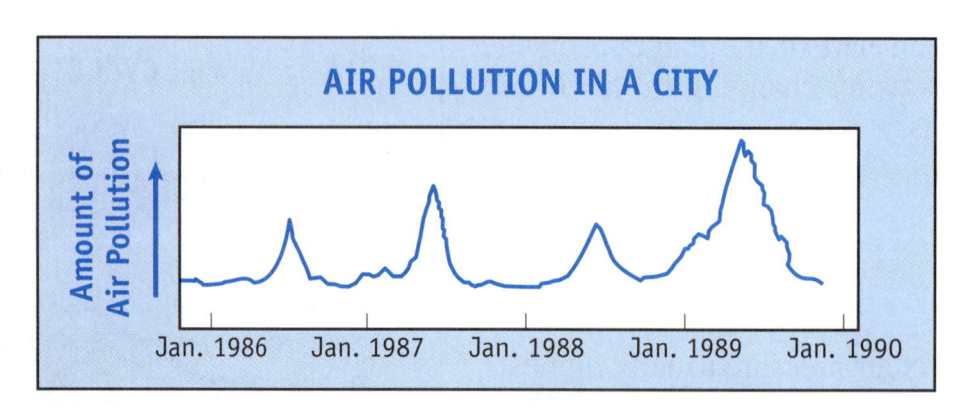

8. The graph above shows the amount of air pollution surrounding a city during four years. Which conclusion is supported by the graph?
   A. The pollution is decreasing at a constant rate.
   B. The pollution data shows no evident pattern.
   C. The pollution is increasing at a constant rate.
   D. The pollution increases sharply each summer.

   SI: B
   G8.3

9. Which is an example of condensation in the water cycle?
   A. Water drops fall through the air.
   B. Streams flow into rivers and rivers into oceans.
   C. Puddles of water quickly disappear on a hot day.
   D. Clouds of water droplets form in the atmosphere.

   ES: C
   G7.3

10. Why are carbon dioxide and water vapor known as the "greenhouse gases"?
    A. They trap thermal energy in the atmosphere.
    B. They are good reflectors of the sun's radiation.
    C. They are found in mixed amounts in Earth's lithosphere.
    D. They are found in varying amounts in Earth's atmosphere.

    ES: C
    G7.2

11. Earth has the ability to absorb and recycle certain materials naturally. This has a significant impact on the environment.

    In your **Answer Document**, explain how Earth naturally recycles carbon dioxide.

    Then explain what the graph shows about Earth's ability to recycle carbon dioxide. (2 points)

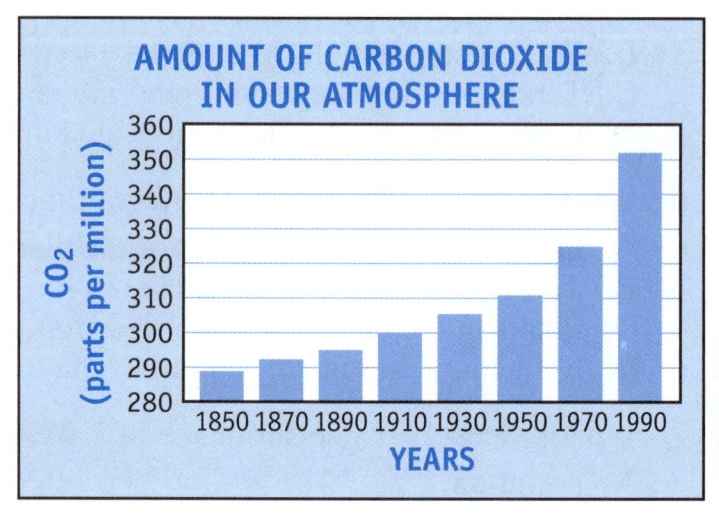

    AMOUNT OF CARBON DIOXIDE IN OUR ATMOSPHERE

    ES: C
    G7.2

CHAPTER 15: THE INTERACTION OF EARTH'S SYSTEMS

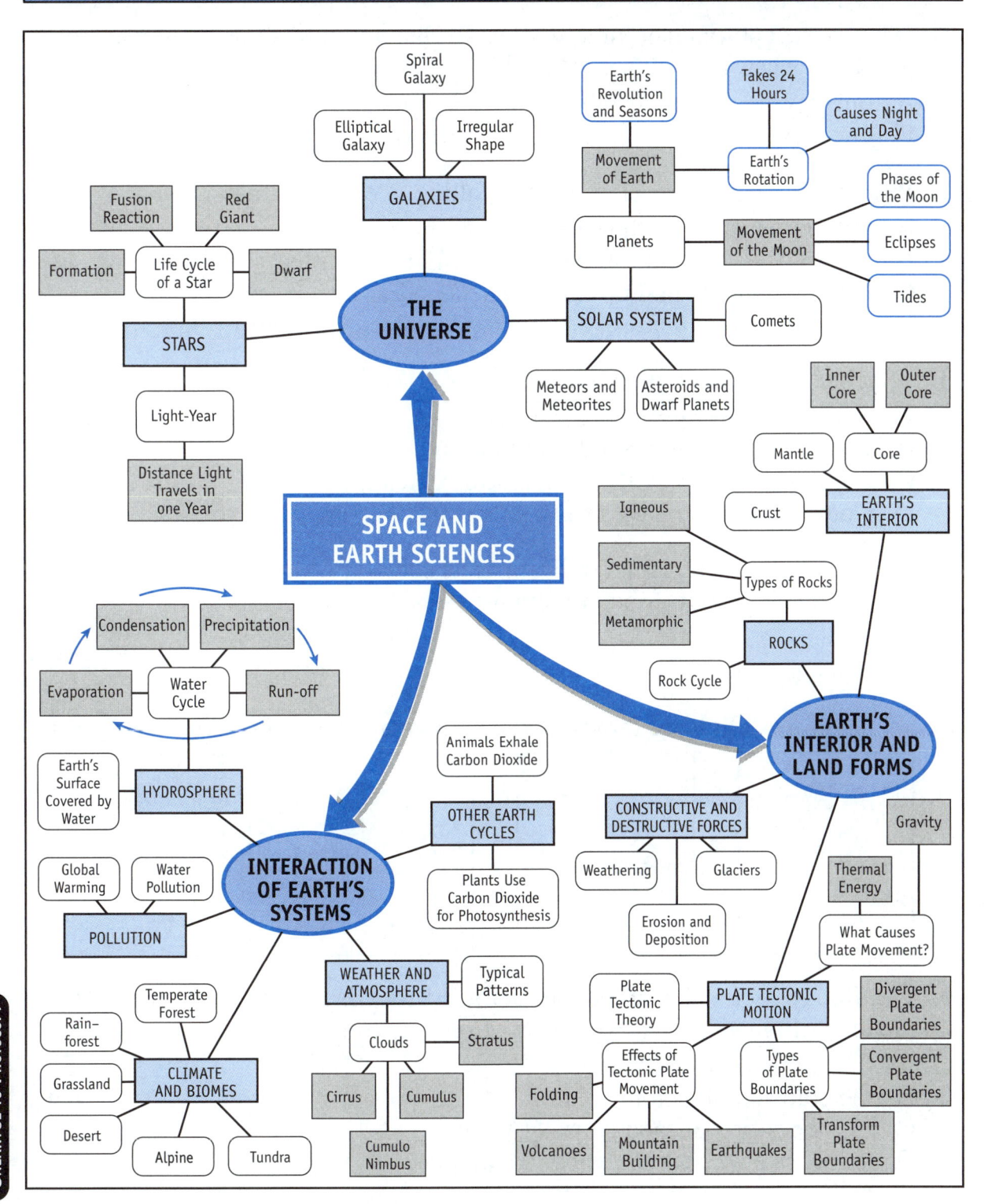

CONCEPT MAP OF EARTH & SPACE SCIENCE

SPACE AND EARTH SCIENCES

- **THE UNIVERSE**
  - GALAXIES
    - Spiral Galaxy
    - Elliptical Galaxy
    - Irregular Shape
  - STARS
    - Life Cycle of a Star
      - Formation
      - Fusion Reaction
      - Red Giant
      - Dwarf
    - Light-Year
      - Distance Light Travels in one Year
  - SOLAR SYSTEM
    - Movement of Earth
      - Earth's Revolution and Seasons
      - Earth's Rotation
        - Takes 24 Hours
        - Causes Night and Day
    - Planets
    - Movement of the Moon
      - Phases of the Moon
      - Eclipses
      - Tides
    - Comets
    - Meteors and Meteorites
    - Asteroids and Dwarf Planets

- **EARTH'S INTERIOR AND LAND FORMS**
  - EARTH'S INTERIOR
    - Core
      - Inner Core
      - Outer Core
    - Mantle
    - Crust
  - ROCKS
    - Types of Rocks
      - Igneous
      - Sedimentary
      - Metamorphic
    - Rock Cycle
  - CONSTRUCTIVE AND DESTRUCTIVE FORCES
    - Weathering
    - Glaciers
    - Erosion and Deposition
  - PLATE TECTONIC MOTION
    - Plate Tectonic Theory
    - What Causes Plate Movement?
      - Gravity
      - Thermal Energy
    - Types of Plate Boundaries
      - Divergent Plate Boundaries
      - Convergent Plate Boundaries
      - Transform Plate Boundaries
    - Effects of Tectonic Plate Movement
      - Folding
      - Volcanoes
      - Mountain Building
      - Earthquakes

- **INTERACTION OF EARTH'S SYSTEMS**
  - HYDROSPHERE
    - Water Cycle
      - Condensation
      - Precipitation
      - Evaporation
      - Run-off
    - Earth's Surface Covered by Water
  - POLLUTION
    - Global Warming
    - Water Pollution
  - OTHER EARTH CYCLES
    - Animals Exhale Carbon Dioxide
    - Plants Use Carbon Dioxide for Photosynthesis
  - WEATHER AND ATMOSPHERE
    - Typical Patterns
    - Clouds
      - Stratus
      - Cirrus
      - Cumulus
      - Cumulo Nimbus
  - CLIMATE AND BIOMES
    - Rainforest
    - Grassland
    - Desert
    - Temperate Forest
    - Alpine
    - Tundra

TESTING YOUR UNDERSTANDING

1. What is the approximate time it takes for the moon to show a complete cycle of phases when viewed from Earth?
   A. one day
   B. one week
   C. one month
   D. one year

2. What is the main cause of rising and falling tides on Earth?
   A. light reflected from the surface of the moon
   B. the gravitational pull of the moon on Earth's oceans
   C. the gravitational pull of other planets on Earth's oceans
   D. the attraction of comets, asteroids, and meteors to Earth

3. Which is an effect of gravitational force?
   A. Stars create energy by nuclear fusion.
   B. Planets revolve in orbits around the sun.
   C. Light travels almost 10 trillion km in a year.
   D. Comets appear to have bright tails when they approach the sun.

4. Scientists refer to the center of Earth as its core. Of which material do scientists believe much of Earth's outer and inner core are both composed?
   A. iron
   B. igneous rock
   C. liquid magma
   D. metamorphic rock

5. Which of the following shows how an elliptical galaxy appears?

   A.   B.   C.   D.

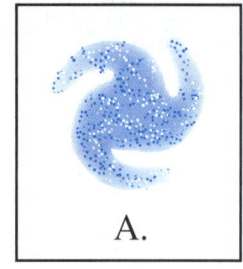

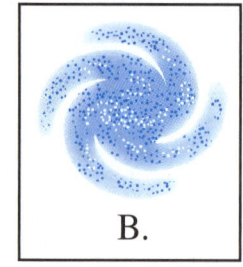

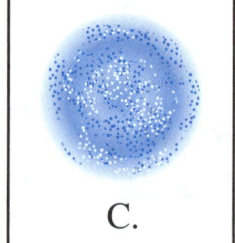

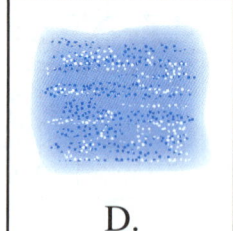

6. The Rocky Mountains are made of folded rock layers. For which method of mountain formation do these layers of rock provide evidence?
   A. the eruption of volcanoes
   B. the collision of tectonic plates
   C. the deposition of sand carried by wind
   D. an ocean plate sinking below a land plate

Scientists classify rocks as igneous, sedimentary, or metamorphic.

Igneous   Sedimentary   Metamorphic

7. What is this classification of rocks based on?
   A. texture
   B. crystal or grain size
   C. method of formation
   D. mineral composition

8. What type of plate boundary is illustrated in the diagram to the right?
   A. transform
   B. divergent
   C. volcanic
   D. convergent

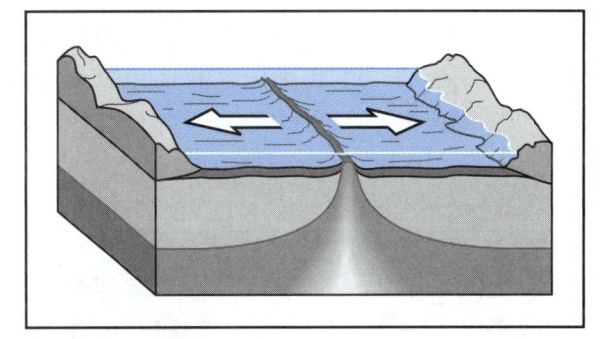

9. Clouds often form after moist air rises. What then happens to the water vapor?
   A. It expands and transpires.
   B. It cools and condenses.
   C. It expands and warms.
   D. It warms and condenses.

10. A group of scientists wants to explore the atmosphere of Saturn. Under the present state of technology, what would be the best equipment for these scientists to use to obtain the most detailed data about Saturn's atmosphere?
    A. space probe
    B. manned spacecraft
    C. telescope from Earth
    D. satellite orbiting Earth

11. Our solar system consists of the sun, planets, moon, asteroids, and comets.

    In your **Answer Document**, identify one way in which comets and asteroids are similar. Then identify one way they differ. (2 points)

12. The water cycle is a key factor in most weather-related events that we experience on a daily basis.

    In your **Answer Document**, describe two weather-related phenomena. Then explain how each of these is connected to what occurs in the water cycle. (4 points)

Use the following chart to answer questions 13 and 14.

Mineral	Colors	Hardness	Luster	Cleavage
Calcite	White, tan, or gray	3	Glassy	3 directions
Gold	Yellow	2.5 to 3	Metallic	None
Quartz	Colorless	7	Glassy	None
Pyrite	Yellow	6 to 6.5	Metallic	None

13. A prospector has discovered a yellow metal. What property can he use to determine if it is gold or pyrite (also known as "fool's gold")?
    A. color
    B. luster
    C. hardness
    D. cleavage

14. A scientist discovers rocks that have a glassy, colorless appearance. Based on the chart, what mineral might they contain?
    A. quartz
    B. calcite
    C. gold
    D. pyrite

CHECKLIST OF LIFE SCIENCE BENCHMARKS

☐ You should be able to describe how the positions and motions of the objects in the universe cause predictable and cyclic events. [Chapter 13]

☐ You should be able to explain that the universe is composed of vast amounts of matter, most of which is at incomprehensible distances and held together by gravitational force. [Chapter 13]

☐ You should be able to describe how the universe is studied by the use of equipment such as telescopes, probes, satellites and spacecraft. [Chapter 13]

☐ You should be able to describe interactions of matter and energy throughout the lithosphere, hydrosphere and atmosphere (e.g., water cycle, weather and pollution). [Chapter 15]

☐ You should be able to identify that the lithosphere contains rocks and minerals and that minerals make up rocks. [Chapter 14 and Chapter 15]

☐ You should be able to describe how rocks and minerals are formed and/or classified. [Chapter 14 and Chapter 15]

☐ You should be able to describe the processes that contribute to the continuous changing of Earth's surface (e.g., earthquakes, volcanic eruptions, erosion, mountain building and lithospheric plate movements). [Chapter 14 and Chapter 15]

UNIT 6

A PRACTICE GRADE 8 SCIENCE ACHIEVEMENT TEST

This unit consists of a practice **Grade 8 Science Achievement Test** — 32 multiple-choice questions, four short-answer, and two extended-response questions. Remember to use the "E-R-A" approach: **Examine** the question to see what is asked. **Recall** what you learned about that particular topic. Finally, **Apply** your knowledge to answer the question. Each question is identified by its standard, benchmark, and grade level indicator. This will help you identify topics you may still need to study. *Good luck on this practice test*!

Use the information below to answer questions 1 and 2.

A science class is exploring a rocky hillside for fossils. They find some fossils from ancient sea-shells. The fossils are located in layers of rock formed out of compacted particles of sand.

1. Which kind of rock are the students investigating?
   A. igneous rock
   B. volcanic rock
   C. sedimentary rock
   D. metamorphic rock

   ES: D G6.1

2. The species of animal that created the fossils is now extinct. What conclusion can be drawn from this fact?
   A. The species migrated to a new location.
   B. The species was unable to adapt to environmental change.
   C. The species successfully adapted to environmental changes in the area.
   D. The local environment changed slowly enough for other species to adapt.

   LS: D G8.5

3. Most fish in Lake Erie reproduce sexually, but some species reproduce asexually. If the temperature of the lake changes because of "global warming," some scientists predict the sexually reproducing fish will survive better than asexually reproducing species. Which evidence best supports this hypothesis?
   A. Asexually reproducing fish multiply more quickly.
   B. Asexually reproducing species promote genetic continuity.
   C. Sexually reproducing fish only mate in specific conditions.
   D. Sexually reproducing species have greater variation in inherited traits.

   LS: B G8.2

223

4. José conducts an experiment to see how the heat from a 60-watt light bulb affects the evaporation rate of water. He places 500 mL of water into three identical beakers. Then he places each beaker at a different distance from the light bulb. José measures the distance of each beaker from the light bulb in centimeters. He then puts the light bulb on for exactly three hours. At the end of this time, he measures the volume of water remaining in each beaker.

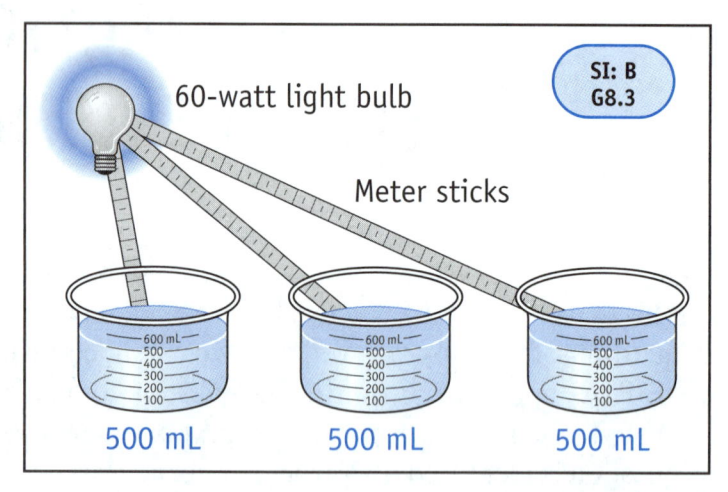

Which would be the best way of graphing the variables in this experiment?

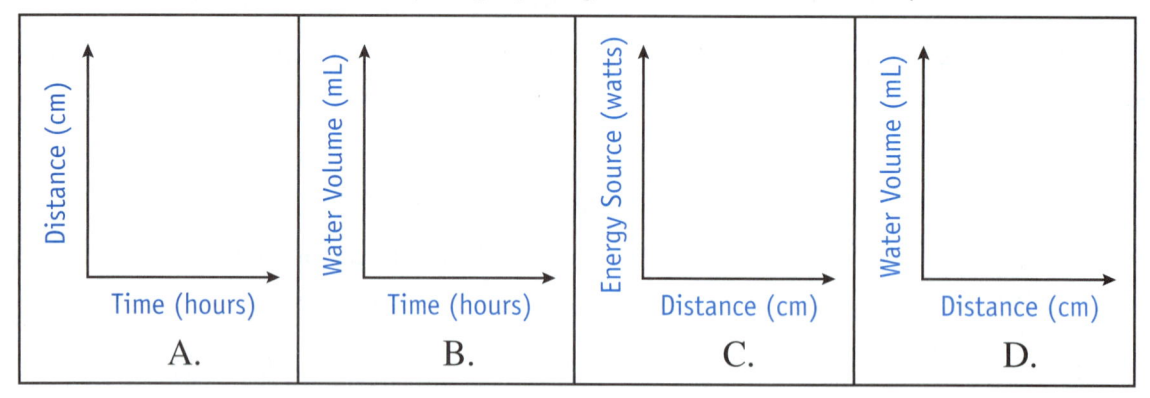

A. Distance (cm) vs. Time (hours)
B. Water Volume (mL) vs. Time (hours)
C. Energy Source (watts) vs. Distance (cm)
D. Water Volume (mL) vs. Distance (cm)

5. A class investigating the properties of different substances collected the data shown in the table below.

Substance	Mass	Volume
Water	2 g	2 cm^3
Copper	17 g	2 cm^3
Aluminum	11 g	4 cm^3
Iron	31 g	4 cm^3

Based on the table, which of these substances has the greatest density?

A. iron
B. water
C. copper
D. aluminum

6. The number of organisms that an ecosystem can support will depend on availability of both biotic (living) and abiotic (non-living) resources.

In your **Answer Document**, describe one abiotic resource and one biotic resource in an ecosystem. (2 points).

7. What is the most important reason for scientists to keep accurate records of their experiments, including the procedures they have performed?
   A. Other scientists can attempt to perform the same experiment better.
   B. Scientists find that disproving a hypothesis can be as important as proving it.
   C. Other scientists are able to repeat the experiment, reducing the risk of bias.
   D. People from other backgrounds may want to conduct similar experiments on related topics.

8. The part of the moon that is lit up each night repeats itself in a cycle. What is the approximate time it takes for the moon to show a complete cycle of phases when viewed from Earth?
   A. 1 day
   B. 1 week
   C. 1 year
   D. 1 month

The diagram below shows the relationship of different levels of organization in a living organism.

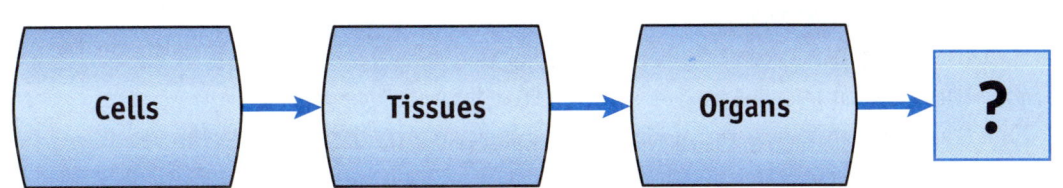

Cells → Tissues → Organs → ?

9. Which of the following best completes the empty box in the diagram?
   A. Habitats
   B. Body Types
   C. Ecosystems
   D. Organ Systems

10. Which event is an example of a chemical change?
    A. A kettle of water boils.
    B. A candle burns on a birthday cake.
    C. A spoon of sugar dissolves in water.
    D. A magnet separates iron filings from sand.

11. There are a wide variety of body plans and internal structures found in multicellular organisms. Which best describes the body plan of the maple tree shown on the right?
    A. radial
    B. unicellular
    C. nonvascular
    D. symmetrical

Use the following information to answer question 12.

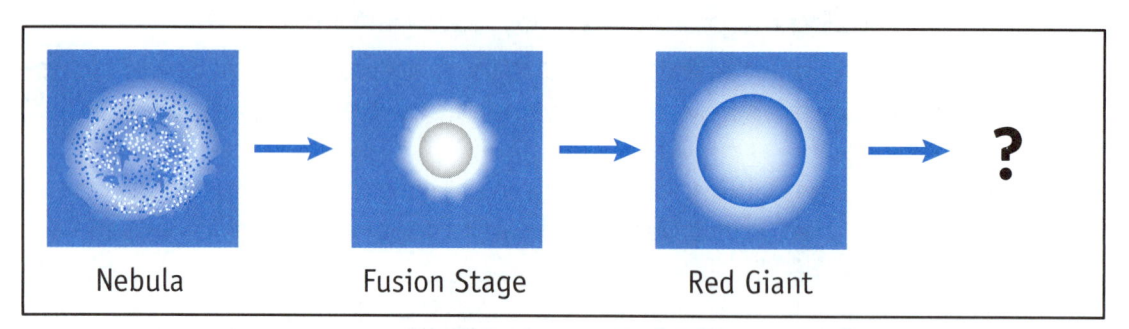

Nebula → Fusion Stage → Red Giant → ?

12. What is most likely to be the next stage in the life cycle of this star?
    A. comet
    B. galaxy
    C. dwarf
    D. planet

    ES: B
    G8.7

13. In the diagram to the right, an unbalanced force is applied to an object at rest. What effect does this have on the object?

    Object at rest

    A. The object remains at rest.
    B. The object makes a circular motion.
    C. The object moves in the same direction as the force.
    D. The object moves in a direction opposite to that of the force.

    PS: B
    G8.3

14. Soil organisms, such as fungi, worms and bacteria, are all part of a woodland ecosystem. Which best describes the role that these organisms play in the woodland ecosystem?

    A. They provide nitrogen gas to animals.
    B. They obtain dissolved oxygen from moisture.
    C. They break down the remains of other living things.
    D. They store chlorophyll for photosynthesis.

    LS: C
    G7.2

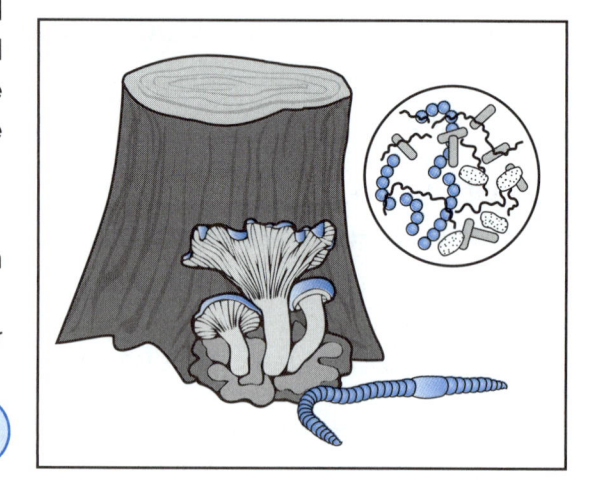

15. How are seismic (earthquake) waves and sound waves alike?
    A. Both are caused by radiant energy.
    B. Both can be heard by the human ear.
    C. Both are caused by chemical reactions.
    D. Both are caused by vibrations in materials.

    PS: D
    G8.5

Use the information below to answer questions 16 to 18.

An ecosystem contains a community of organisms living in one area. The diagram below illustrates a food web in a pond ecosystem. The ecosystem includes the pond and pond grass growing inside and around the pond.

16. What would most likely occur in this pond ecosystem if a new type of frog were introduced that ate mosquito larvae?

    A. The number of yellow perch would increase.
    B. The amount of algae and pond grass would decrease.
    C. The amount of algae and pond grass would increase.
    D. The number of valve snails would decrease.

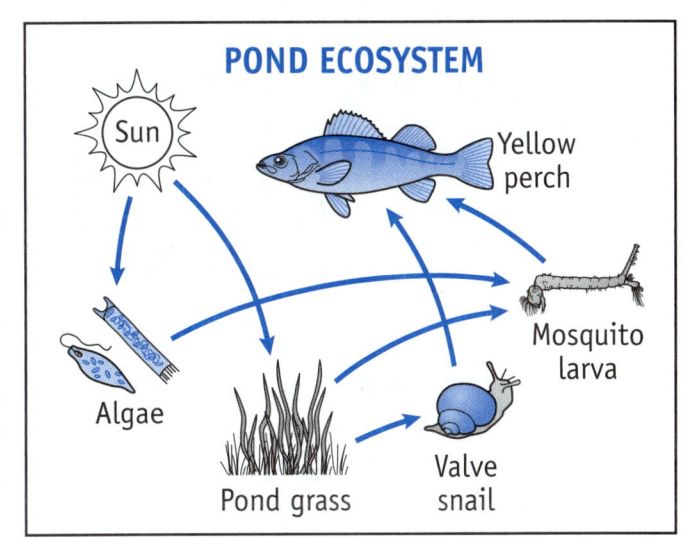

POND ECOSYSTEM (Sun, Yellow perch, Mosquito larva, Algae, Pond grass, Valve snail)

17. Based on information in the diagram, how could the yellow perch be described?

    A. prey
    B. host
    C. parasite
    D. predator

18. What process allows the algae and pond grass to convert radiant energy from sunlight into chemical energy?

    A. digestion
    B. respiration
    C. photosynthesis
    D. cellular respiration

19. Men and women from all countries and cultures have made important contributions to the growth of scientific knowledge.

    In your **Answer Document**, identify two scientists. Then describe one contribution each of these individuals has made to the progress of science. (4 points).

20. What is an important effect of tectonic plate movement?

    A. the recycling of organic compounds
    B. frequent tropical storms above lava flows
    C. the formation of mountains where land plates collide
    D. the weathering and erosion of landforms where plates separate

A scientist places 6 bean plants in different locations in his laboratory. He gives each plant the same quantity of food and water. However, he also gives 50g of a commercial fertilizer to the three plants in Group I. He then measures the heights of the plants at the end of a six-week period. Below are the heights of the plants:

GROUP I: WITH FERTILIZER

Plant	Height
Plant A	70 cm
Plant B	80 cm
Plant C	80 cm

GROUP II: WITHOUT FERTILIZER

Plant	Height
Plant A	60 cm
Plant B	50 cm
Plant C	70 cm

21. What hypothesis is the scientist testing in this experiment?
    A. The location of a bean plant will have an effect on its growth.
    B. Commercial fertilizers are too costly to use with certain plants.
    C. A bean plant will grow more with fertilizer than without fertilizer.
    D. Bean plants grow the same when given the same amount of water and food.

22. What is the median height of these six bean plants?
    A. 50 cm
    B. 60 cm
    C. 70 cm
    D. 80 cm

Use the weather map below to answer question 23.

23. Based on the map to the right, what weather is most likely for Austin in the next several days?
    A. warmer and drier
    B. cooler and clearer
    C. warmer and wetter
    D. cooler and cloudier

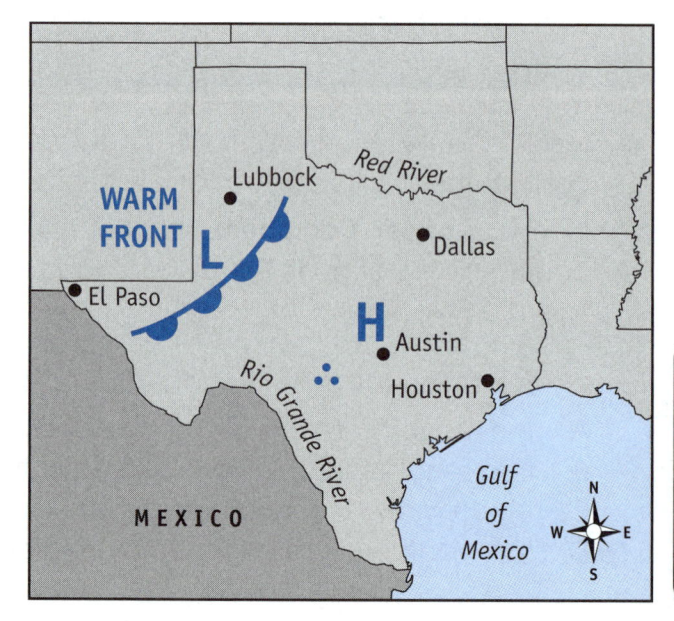

Use the information in the diagram to answer questions 24 and 25.

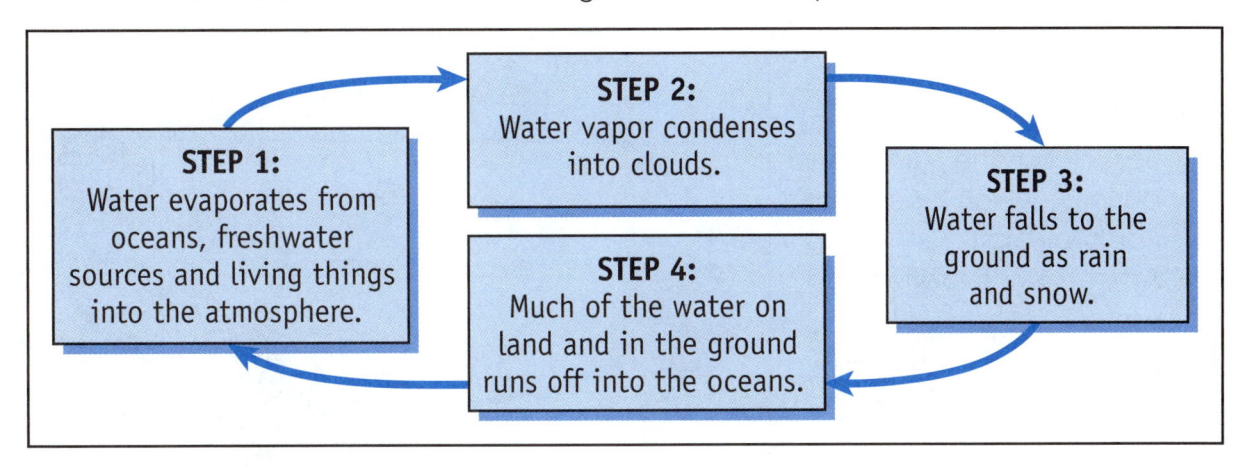

STEP 1: Water evaporates from oceans, freshwater sources and living things into the atmosphere.

STEP 2: Water vapor condenses into clouds.

STEP 3: Water falls to the ground as rain and snow.

STEP 4: Much of the water on land and in the ground runs off into the oceans.

24. Which step of the water cycle makes surrounding areas cooler because it absorbs energy?
    A. Step 1
    B. Step 2
    C. Step 3
    D. Step 4

    ES: C
    G7.3

25. What would be the most likely effect on the water cycle if Earth's average temperatures rose because of "global warming"?
    A. less evaporation and less precipitation
    B. greater evaporation and less precipitation
    C. less evaporation and greater precipitation
    D. greater evaporation and greater precipitation

    ES: C
    G7.3

26. A student has learned about kinetic and potential energy in her science class. She decides to conduct a simple experiment at her home. She opens the window of her second floor bedroom and drops a beach ball from the window. What describes what happens as the ball falls from the window?

    A. Its kinetic and potential energy both decrease.
    B. Its kinetic energy and potential energy both increase.
    C. Its kinetic energy increases as its potential energy increases.
    D. Its kinetic energy increases as its potential energy decreases.

    PS: D
    G7.2

27. As shown in the diagram to the right, Earth's axis is tilted as it orbits the sun. Based on the diagram, during which part of Earth's orbit is North America experiencing winter?

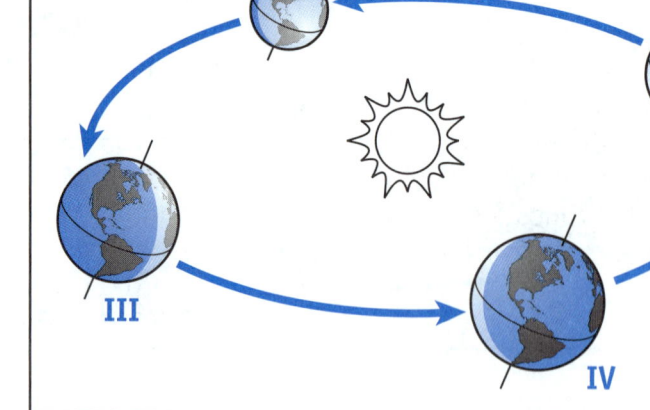

A. I
B. II
C. III
D. IV

ES: A
G8.1

28. Why does North America experience colder temperatures during winter?

A. Earth is farther from the sun in winter than in summer.
B. Gravity bends the sun's rays away from North America.
C. Greater amounts of water vapor in the atmosphere block sunlight.
D. Sunlight does not hit North America as directly as during the rest of year.

ES: A
G8.1

29. A scientist decides to conduct an experiment using a mixture. He begins by dissolving table salt (NaCl) in a glass of water. How does the scientist know that a physical change, not a chemical change, has occurred?

A. The salt is no longer visible.
B. The salt cannot be separated from the water with a filter.
C. The mixture has a different density than the water without salt.
D. The chemical properties of the salt and water have not changed.

PS: A
G6.3

30. Reproduction through either sexual or asexual reproduction is necessary for the continuation of each species. Which is an example of asexual reproduction?

A. A young chicken is not identical to either of its parents.
B. An egg and sperm combine to produce a new organism.
C. A fragment of an earthworm grows into an entire organism.
D. Two mice each contribute some inherited traits to their offspring.

LS: B
G6.5

31. Technology plays an important role in meeting human needs and improving the quality of life for many people.

    In your **Answer Document**, identify one need that has been met by technology.

    Then describe one consequence this technology has had in addition to solving the problem. (2 points)

32. Which example shows a change from chemical energy to light energy?
    A. A candle burns with a flame.
    B. A car engine ignites gasoline to move its cylinders.
    C. A hydroelectric plant uses falling water to turn its turbines.
    D. An electric lamp is plugged into a wall socket and turned on.

33. Sam takes a trip to visit his relatives. The chart below shows the total distance traveled from the starting point after each hour of the trip.

Time	Distance Traveled
1 hour	50 miles
2 hours	100 miles
3 hours	150 miles
4 hours	200 miles

    In your **Answer Document**, create a line graph showing the time spent and distance traveled for all four hours of the trip. Label the X and Y axis (2 points).

34. The surface of Earth undergoes constant change. Some of the forces changing Earth's surface are constructive and others are destructive.

    In your **Answer Document**, describe one constructive and one destructive force that shapes Earth's surface. (2 points)

35. A meteorologist is studying Earth's atmosphere in order to predict the weather. The meteorologist observes gathering cumulonimbus clouds and falling barometric pressure. What type of weather should she expect?
    A. cold weather
    B. heavy, dense fog
    C. rain and thunder
    D. warm and sunny weather

36. A scientist rolls a ball down a hill. The ball rolls a distance of 50 m in 5 seconds.

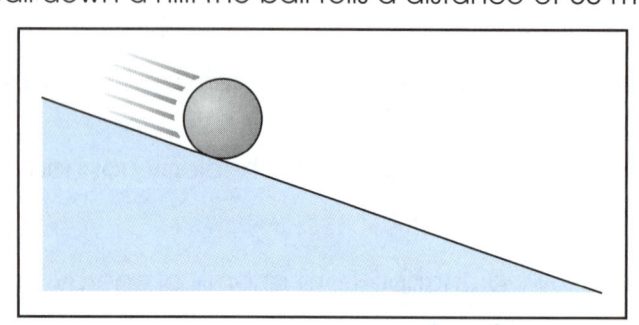

    What is the speed of the ball?
    A. 5 m/s
    B. 10 m/s
    C. 50 m/s
    D. 100 m/s

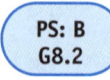

37. Renewable energy sources are important because they are available indefinitely. Many do not produce the "greenhouse" gases that most nonrenewable sources of energy produce.

    In your **Answer Document**, describe one renewable energy source and one nonrenewable energy source.

    Then explain how each one can be used to create electricity (4 points).

Use the Venn Diagram below to answer question 38.

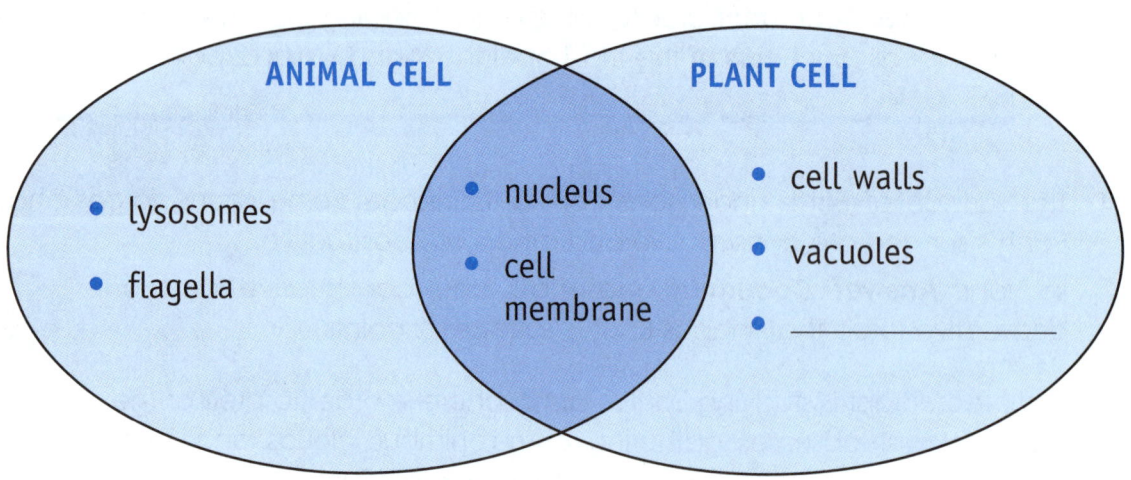

ANIMAL CELL
- lysosomes
- flagella

(overlap)
- nucleus
- cell membrane

PLANT CELL
- cell walls
- vacuoles
- _____

38. What term is missing from the blank line on the Venn diagram?
    A. proteins
    B. ribosomes
    C. chloroplasts
    D. nuclear membrane

GLOSSARY / INDEX

The number in brackets indicates the page number where the term is first discussed.

Abiotic Resource. Any nonliving resource in an ecosystem, such as water. [159]
Adaptation. Processes by which species survive environmental change. [161]
Animal. A type of multicellular organism that can move freely but cannot produce its own food. An animal eats plants or other animals to survive. [135–136]
Asexual Reproduction. Method of reproduction in which an organism can duplicate itself; allows rapid reproduction and insures genetic continuity. [146]
Asteroid. Pieces of rock and metal orbiting the sun; many are found in the "asteroid belt" between Mars and Jupiter. [181]
Atmosphere. Envelope of gases, mainly nitrogen and oxygen, surrounding Earth. [210]
Atom. Smallest unit of any element. Atoms contain protons, neutrons, and electrons. [85]
Automation. Replacement of human labor by machines, especially in factories. [68–69]

Bias. A prejudice that prevents a person from viewing events impartially. [30]
Biomes. An area where, because of climate and resources, particular plants and animals live: for example, there are desert, alpine, grassland, forest, and tundra biomes. [214]
Biotic Resource. Any living resource in an ecosystem. [155]

Cell. The basic unit of all living things. Cells reproduce themselves. All cells have DNA, a cell membrane, and cellular fluids. [128]
Cellular Respiration. Process of obtaining energy by combining oxygen with glucose. [130]
Cell Wall. Stiff outer wall surrounding plant cells. [130]
Chemical Change. When a substance changes its structure and chemical properties by combining with another substance or by separating into new substances. [93]
Chemical Property. The ability of a substance to combine with others in a chemical reaction. [87]
Chloroplasts. The internal structures of plant cells that conduct photosynthesis. [131]
Conservation of Energy. Energy can be transformed but not destroyed. [112]
Conservation of Matter. Matter can change in a reaction but cannot be destroyed. [93]
Constraint. A limit that is imposed on a technological project or design. [72–73]
Consumer. Any organism that must eat other organisms to survive. [157]
Commensalism. Relationship where one organism benefits without hurting the other. [156]
Core. The center of Earth. The outer core is liquid iron and nickel; the inner core is solid iron. [193]

Decomposers. Organisms in an ecosystem, such as ants, worms, and bacteria, that live by breaking down waste products and dead organisms. [157]
Density. The mass of a substance per unit of volume (mass/volume); for example, g/cm^3. [89]

Description. An account of events that describes or tells what happens. [26]
DNA. The molecular basis of heredity. [128]

Earthquake. Vibrations of Earth's crust that send out seismic waves. [197]
Eclipse. Event occurring when the moon blocks the sun's light (solar eclipse) or when the moon passes into Earth's shadow (lunar eclipse). [185]
Ecosystem. All the living and non-living things in an area; they affect and depend on each other. Energy in an ecosystem flows from producers to consumers. [154]
Electricity. A form of energy made by fast-moving particles. [110]
Energy. The ability to do work. Energy takes different forms: electricity, light, and thermal energy are all different forms of energy. [107–112]
Erosion. The process by which rock and soil are worn down and carried away by wind, running water, ocean waves or glaciers. [199]
Explanation. An attempt to explain why or how something happened. [26]
Extinction. Process in which every member of a species dies out. Extinction occurs when a species fails to adapt to changed conditions: for example, dinosaurs are extinct. [164]

Folding. A process in which Earth's tectonic plates are pushed together to create mountains. [196]
Force. A push or pull acting on an object. An **unbalanced force** will cause it to move or change its motion. [102–103]
Fossil. Impression left in sedimentary rock by the remains of a dead plant or animal. [164]
Front. Where a mass of warm air meets colder air (warm front) or where a mass of cold air meets warmer air (cold front). [213]

Genetic Continuity. The continuation of genes, passed from parents to offspring. [146]
Global Warming. Increased carbon dioxide in the atmosphere acts as a blanket, trapping in heat. This has led to higher temperatures across the globe. [70, 210]
Gravity. A non-contact force of attraction between objects. Gravity pulls objects to Earth and influences the motion of bodies in outer space. [180]

Hurricane. A type of storm with severe winds, when ocean waters become hot in tropical regions. [211]
Hydrosphere. All of the water on Earth. [207]
Hypothesis. Educated guess that attempts to answer a question and can be tested. [28]

Igneous Rock. Rocks formed from lava that has cooled. [201]
Inherited Trait. A fixed characteristic of an organism determined by genes that are inherited from its parent or parents. [143–144]

Kinetic Energy. Energy of movement. Any moving object has kinetic energy. [108]

Light. A form of radiant energy that can travel through some materials or space. [110]
Light-Year. Unit of distance based on how far light travels in one year. [176]
Lithosphere. Land surface and crust of Earth. The lithosphere consists of shifting tectonic plates. [207]

Mantle. Largest section of Earth's interior, made of hot, semi-liquid rock. [193]
Mass. The amount of matter something has, measured in grams or kilograms. [89]
Matter. Anything that has mass and takes up space. [84-85]
Mean. The average value in a set of numerical results. [56]
Median. The middle value in a set of numerical results. [56]
Metamorphic Rock. Rocks that have been changed into another kind of rock by heat and pressure under Earth's surface. [201]
Meteor. Piece of rock or metal from space that falls into a planet's atmosphere. A **meteorite** is what is left when the meteor reaches the planet's surface. [181]
Mode. The most common value in a set of numerical results. [56]
Mountain-building. When two tectonic plates collide, Earth's crust sometimes folds. This pushes some of Earth's crust upwards and creates mountains. [197]
Motion. When an object changes its position over time. Motion consists of both speed and direction. Motion must be measured with respect to a reference point. [100]
Multicellular. Organism made up of many cells. [132]
Mutualism. Relationship between organisms where both benefit; for example, bees spread a plant's pollen to other plants, benefiting both the bees and the plants. [156]

Nonrenewable Energy Source. An energy source, like fossil fuels, that can be used up. [115]
Nuclear Energy. Energy released by the fission or splitting apart of atomic nuclei. [110]

Organism. A living thing. All organisms are made of one or more cells and can reproduce. [128]
Organ and Organ System. A group of tissues acting together to perform a function, like the human digestive system. [133]

Parasite. An animal that lives off another, known as its **host**, without immediately killing it. [150]
Photosynthesis. The process by which plants capture energy from sunlight and convert it into a form of sugar: water + carbon dioxide + light $\longrightarrow$ glucose + oxygen [131]
Physical Change. When an object changes one or more of its physical properties. For example, a physical change occurs when a pot of water boils. [92]
Physical Property. Characteristic of a substance, such as mass, state, or appearance. [87]
Planet. Large bodies of rock, often with ice or gas, that orbit the sun and dominate their orbits: Mercury, Venus, Earth, Mars, Jupiter, Saturn, Uranus, and Neptune. [179]
Plant. A multicellular organism with cell walls, vacuoles, and chloroplasts. Plants cannot move from place to place, but produce their own food through photosynthesis. [130–131]
Potential Energy. Energy stored in an object, such as a spring or chemical bond. [109]
Predator. An animal that hunts and eats other animals, which are known as **prey**. [155]
Producer. A living thing that can make its own food; a plant. [157]

Radiant Energy. Energy of light or other radiomagnetic waves. [110]
Renewable Energy Source. A source of energy that can last indefinitely, such as wind, waterpower, sunlight, or biomass. [116–117]
Rock. A solid made of minerals found on Earth's surface or below it. [200]
Rock Cycle. Rocks change form between igneous, sedimentary, and metamorphic rocks. [202]

Scientific Inquiry. The way scientists approach the natural world: usually begins with observation of the natural world, followed by questions about what is observed. [40]

Sedimentary Rock. Rock made by layers of sand, mud or other materials that are deposited and pressed together. Fossils are often found in sedimentary rock. [201]

Sexual Reproduction. Method of reproduction in which two organisms each contribute one half of the genes needed to produce a new offspring; increases genetic variation. [147]

Skepticism. The quality of asking questions and not just accepting what others say. [33]

Smog. A thick mixture of smoke from pollution and fog. [209]

Star. Enormous ball of super-heated gases in space. Each star goes through a life-cycle: nebula, a mainstream star where fusion is occurring, a red giant, and a dwarf or black hole. [176–177]

Symbiotic Relationship. A relationship between organisms in which each organism depends on the other. [155]

Technology. The use of tools and techniques for making and doing things. [64–66]

Technological Design. The first step is to identify a need; then identify possible solutions; next design a solution; and finally evaluate and test the solution. [71–72]

Tectonic Plate Boundaries. Areas where plates meet or spread apart. Plates collide at **convergent plate boundaries**. Plates spread apart at **divergent plate boundaries**. Plates slide by each other horizontally at **transform plate boundaries**. [194]

Thermal Energy. Energy caused by the movement of particles that make up matter. [108]

Tide. Rise and fall of Earth's oceans, caused by the moon's gravitational attraction. [185]

Tissue. A group of identical cells acting together to perform a single function; such as muscle tissue or bone tissue. [133]

Tornado. A funnel of swirling wind that sucks up buildings and property. [211]

Unicellular. An organism with just one cell, like bacteria. [132]

Variable. Anything that can be changed or that might change in an experiment. An **independent variable** (X axis) is what a scientist deliberately changes or manipulates. The **dependent variable** (Y axis) is what changes or responds as a result. [44]

Volcano. An opening in Earth's surface that lets out molten lava and gases. [198]

Volume. How much space something takes up, often measured in cm^3 or mL. [89]

Water Cycle. A cycle in which water from lakes and oceans is heated by energy from the sun, **evaporates** into the atmosphere, **condenses** into drops of water that float as clouds, falls back to Earth as **precipitation** (rain, snow, or hail) and then **runs off** land areas back into lakes and oceans. [208]

Water Depletion. When lakes and rivers dry up because of a lack of water. [210]

Weather. Conditions in the atmosphere that change daily. [210]

Weathering. The gradual wearing down of rocks on Earth's surface by the actions of the wind, water, ice and living organisms. [198]